Integrating Mental Health and Disability Into Public Health Disaster Preparedness and Response

Integrating Mental Health and Disability Into Public Health Disaster Preparedness and Response

Jill Morrow-Gorton
University of Massachusetts Chan Medical School,
Worcester, MA, United States

Susan Wolf-Fordham
Consultant, Association of University Centers on Disabilities, Silver Spring, MD,
United States
Adjunct Faculty of Public Health, Massachusetts College of Pharmacy and
Health Sciences, Boston, MA, United States

Katherine Snyder
Independent Contributor

ELSEVIER

Butterworth-Heinemann
An imprint of Elsevier

Butterworth-Heinemann is an imprint of Elsevier
The Boulevard, Langford Lane, Kidlington, Oxford OX5 1GB, United Kingdom
50 Hampshire Street, 5th Floor, Cambridge, MA 02139, United States

ISBN: 978-0-12-814009-3

For information on all Butterworth-Heinemann publications
visit our website at https://www.elsevier.com/books-and-journals

Publisher: Candice G. Janco
Acquisitions Editor: Kathryn Eryilmaz
Editorial Project Manager: Aera F. Gariguez
Production Project Manager: Sreejith Viswanathan
Cover Designer: Mark Rogers

Typeset by STRAIVE, India

Dedication

*We dedicate this book collectively and individually to several people.
First, we dedicate the book to our late friend and colleague, Dr. Richard Beinecke, Rick to us. He was integrally involved in the planning and vision of the book, but sadly did not get to see it to fruition. We thank him for his input and have missed his guidance.
Jill dedicates this book to her family, particularly to her daughter Kate who while homeschooling her children during the pandemic drafted the law chapter and to her husband Kit whose understanding and assistance were immeasurable.
Sue dedicates this book to Pete, Lisa, Danny, Esther, and Tzvi Fordham and to Judy and Bob Wolf. Lisa's perseverance, tenacity, feistiness, and sense of humor in the face of exclusion, ignorance, and disability bias are an inspiration.
Kate dedicates this book to her mom, her husband Brian, and Cricket, Cricky for short, who is a diligent foot warmer.*

Contents

Preface

The goal of this book is to bring together the fields of mental health, chronic illness, and disability (collectively referred to as disability) and public health emergency preparedness and response in order to draw attention to the disaster needs of people with disabilities and the opportunities of disaster-inclusive practices. We face various disasters. They include natural disasters such as earthquakes, storms and floods, and human-caused disasters such as bombings, terrorist attacks, hazardous chemical spills, and pandemics. We ask the question of what we can learn from the lived expertise, strengths, and disaster experiences of people with disabilities that will demonstrate how to better partner with them to meet their needs in planning and responding to a disaster, providing services during a disaster, and recovery in the months and years after a disaster.

Children and adults with disabilities, mental or behavioral health conditions, and chronic illness remain more vulnerable to the negative effects of emergencies and disasters than the general population. Disability advocates, scholars, and practitioners have written much about the negative impacts of disasters on these populations. For too long, people with disabilities have been left out of the emergency planning process. Local, state, and federal emergency plans omitted specific actions to address their access and functional needs, indicating these needs may not have been considered during the planning process leaving public health and safety emergency responders, planners, and managers unaware or unsure of how to address disability-related needs during and after emergencies and the legal obligation to do so. The consequences of these actions leave those people with disabilities who are unable to evacuate without assistance waiting for help, sometimes in vain. These consequences contribute to slower disaster recovery, and sometimes failure to achieve full recovery, and cause additional disability in the form of physical injury or psychological trauma. In all of this, we need to capture the valuable lessons and experience of people with disabilities who problem-solve daily about how to navigate their social and physical environments.

Multiple attempts to redress the situation are ongoing. The United Nations, through a series of frameworks developed at world conferences, has set forth priorities for disability-inclusive planning practices. Governments, including the US government, advocate approaching disaster planning from a whole-community standpoint, incorporating all of the community including people with disabilities and other potentially vulnerable populations. Guidance for including people with disabilities in emergency planning has come from all levels of government. Disability advocacy organizations have disseminated reports and worked collaboratively at local, state, and federal levels to address these gaps in emergency planning and response. Despite these attempts, gaps in knowledge, experience, and practice remain.

This book attempts to fill some of those gaps by bringing together three fields that traditionally have functioned separately in terms of both knowledge and practice in the face of disaster. It addresses the effects of emotional trauma, personal growth and resilience, the impact on physical health and systems of care, legal compliance, and advocacy with and for people with disabilities before, during, and after disasters. The book also brings the power of community and universal design as strategies to incorporate people with disabilities to meet their own and the community's needs. The book further describes and offers solutions and promising practices for addressing disability and disaster needs, planning, and systems for service delivery at multiple levels: individual, local, state, and federal. It seeks to teach principles and practical strategies for implementing a whole-community and functional and access needs approach and to identify ways to build community and individual resilience to reduce disaster risk for people with disabilities. We hope that this book will encourage and continue the conversation between people with disabilities, practitioners, and academics in each of these fields.

As public health plays an important role not only in acute health issues but also in chronic conditions, the book examines disasters and the inclusion and integration of people with disabilities and chronic physical and mental health conditions through the lens of both public health and emergency management. It begins with an introduction to emergency management history, models, and background epidemiologic and functional information about people with chronic physical and mental health conditions and their intersection with disability. It includes the worldview of disaster management and risk reduction as well as the structure of the US emergency management system, whole-community philosophy, and laws behind disability inclusion. As understanding factors underlying vulnerability and the potential for disaster resilience in these populations are key to successful disaster management, both individual and community resilience are discussed, incorporating the factors important to building each. The final chapter presents some promising disability-inclusive disaster management practices from across the world. These help illustrate ways to engage people with disabilities and others with access and functional needs where they live to be able to incorporate them into advanced planning for disasters in order to avoid repeating past occurrences. Reflection questions throughout the chapters allow for application of the principles within each chapter and further exploration of the topics.

Writing this book during the COVID-19 pandemic has highlighted the issues faced by people with disabilities and the disparities that result. People with disabilities and chronic and mental health conditions have been more likely to die of COVID-19, but also, in some places, generally due to strong advocacy, are at the top of the list to receive protective vaccines. In the midst of this, we found multiple inspirations for the book. We began our book-writing journey with Rick Beinecke, who joined us at the very beginning and shared his enthusiasm, sense of humor, dedication, and expertise in mental health, public administration, and research related to the mental health consequences of the Boston Marathon Bombing in 2013. Unfortunately, he was not able to join us as we complete this journey, but we acknowledge his contribution to our work. While we each approach the topic from different perspectives including medicine, law, government, human services, and personal experience, we share a belief in the importance of not only disability-inclusive emergency planning and response but also disability-inclusive communities that provide a richer environment for everyone.

Jill
Sue
Kate

Introduction

Language used in this book
Identity first and person first language

Some people prefer to use "identity-first" language, and say they're "disabled." Other people prefer to identify themselves as a "person with a disability." The latter is an example of "person-first" language.

This text uses both "person first" and "identity first" language.

"Deaf woman" is an example of identity-first language. Many Deaf people consider Deafness to be a culture and part of who they are. Other examples of identity-first language are "autistic person" or "disabled person." Individuals who identify this way would say that their autism or disability is central to who they are.

Person-first language puts the individual at the front of the description, showing that the person is not just their disability. Putting the person first also emphasizes their humanity. For example, a person with Down syndrome may want to be described by other attributes, not only Down syndrome.

As in emergency planning and response, disability is not one-size-fits-all. It's about which language an individual with a disability prefers.

Other language

In this book we generally use the term "disability" to include chronic physical and mental health conditions as well as disability, however we recognize differences among them.

We also often use "emergency management" to include both emergency management and public health preparedness, although we recognize distinctions between the two fields.

Frameworks and models of disaster management theory: Setting the stage

1

Jill Morrow-Gorton

University of Massachusetts Chan Medical School, Worcester, MA, United States

Learning objectives

1. Define "disaster" and describe different types of disasters.
2. Describe the history of evolution of emergency management in the United States over the 20th and 21st centuries.
3. Describe and apply different models of emergency management to varied disasters.

Integrating Mental Health and Disability Into Public Health Disaster Preparedness and Response
https://doi.org/10.1016/B978-0-12-814009-3.00010-6

Introduction

Emergency management encompasses the strategic processes used to protect critical functions of an organization or community from hazards or risks that can cause disasters to ensure continued operation. Historically the purview of government and law enforcement, emergency planning and response traditionally approached disasters and emergencies from the principles of common good with the focus on protection of infrastructure, assets, and the continuity of business operations. Experience shows that this tactic leaves behind the segment of the community who may see, hear, move, think or communicate differently. Movement toward more inclusive practices has occurred with the incorporation of the concept of the "whole community" inclusive emergency management philosophy. However, implementation of this concept has been uneven and fragmented, leading to a continued pattern whereby the members of the community with the fewest resources, such as people with disability, chronic health problems and mental illness, are the most disproportionately impacted by emergencies and disasters. From the vantage points of ethics and justice, all persons share humanness and the common good, and the interests of the community are best met by respecting and considering the rights and well-being of all of its members. Inclusive emergency planning design, by its nature, responds not only to the needs of those with disabilities, but also to the greater community. One illustration of this is installing sidewalk curb cuts, which benefit people pushing children's strollers, older people who may experience limited mobility, as well as those using wheelchairs. Therefore, an inclusive approach that integrates disability, chronic and mental health needs, whether pre-existing or caused by an emergency event, will benefit the whole community and reduce the impact of disasters on its most vulnerable members. The principles and approach to inclusive emergency management and public health preparedness suggested in this book share common, similar strategies for planning and response. They will be illustrated in the context of an "all hazard" or all types of disasters approach, with attention to all disaster phases, preparedness, prevention, response, recovery, mitigation, to address integrating mental health and disability into emergency planning, responding to disasters, and recovering post disaster [1]. In order to build individual, organizational and community resilience, "whole community" and "all hazards" planning and response must integrate public health, mental health, social and disability systems to support and engage all communities and their members, including those with disabilities. By bringing together the principles, practices and collective knowledge of each of these, emergency management and public health preparedness will have greater success in minimizing the impact of disasters on all of society.

To begin this discussion of integrating public health, mental health and disability into emergency management, it is important to understand disasters and emergency management. This introduction walks through accepted definitions for disaster as well as some details about their causes and impacts. It provides a history and evolution of the emergency management field in order to help understand the original underlying principles and practices which shaped today's systems. The chapter also delves into some of the emergency management models and the theory behind those to better appreciate the thinking about the factors and root causes that impact disaster outcomes. It concludes with an introduction to vulnerability and resilience which are important elements of both public safety and public health emergency management.

What is a disaster?

Disasters occur across the world and impact many people. Before delving into the conceptual models of emergency management, it is important to understand what is meant by disaster. In 2007 the

International Federation of Red Cross and Red Crescent Societies (IFRC) [2] defined disaster as "a serious disruption of the functioning of society which poses a significant, widespread threat to human life, health, property or the environment, whether arising from accident, nature or human activity, whether developing suddenly or as the result of long term processes, but excluding armed conflict." The key features of a disaster are: (1) serious disruption of society, (2) widespread threat to human life, property, and the environment, and (3) a cause which may be natural or human-caused. The length of time the disaster is present, the cause or nature of the disaster, and the size of the geographic area involved define potential impact. Disasters may be ultra-short where the event itself occurs within seconds or minutes. The Boston Marathon bombing in 2013 is one example where the actual bomb blasts lasted less than a minute [3]. Earthquakes also may only last seconds to minutes [4]. Storms, including hurricanes and nor'easters, may last a day or more, but in general have relatively short durations [5,6]. Famine represents a longer term disaster and can last from months to years [7]. Often famine follows another disaster that caused the damage leading to the food shortage. Drought, flooding, storms, excessive temperatures (both hot and cold), war, and plant disease destroy crops and cause food shortages and famine. In the 20th and 21st centuries most of the world's famines have been in Africa, including Ethiopia in the 1950s and Biafra, now part of Nigeria, in the 1960s. Another longer term consequence of disasters are infectious disease outbreaks, especially cholera and malaria related to the lack of a clean water source or standing water which encourages mosquito growth after a flood [8]. Infectious diseases can also cause disasters in the absence of a preceding event, with long lasting and more widely spread effects [9]. Pandemics such as the Ebola outbreaks from 2014 to 2016 in Western Africa and the Severe Acute Respiratory Syndrome Coronavirus 2 (SARS-CoV-2 or COVID-19) which became a pandemic in 2020 can also last for a longer period of time [10,11].

Infectious disease

As infectious diseases have the potential for a global impact, it is important to understand the language used to describe these events plus the characteristics of the organisms involved that allow them to proliferate and cause such events. Public health entities and systems such as the Centers for Disease Control and Prevention (CDC) in the United States and the World Health Organization (WHO) manage complex health events caused by infectious diseases and guide the standardization of infectious disease nomenclature to facilitate communication and understanding of these events [12–14]. Important terms for infectious diseases include "endemic," "outbreaks," "epidemics," and "pandemics." "Endemic" infections occur at a low stable rate in the population. "Outbreaks" of infectious diseases are defined as an unexpected increase in the number of cases in a local area. Outbreaks that affect more people in a population or region are referred to as "epidemics." Epidemics that spread across multiple countries or continents become "pandemics," as seen with both Ebola and COVID-19 [9,10].

Infectious diseases may have a differential impact depending on the characteristics of the organism, where they appear and the population that they affect. Not all infectious diseases have the potential to cause outbreaks or lead to pandemics. Organisms that can cause infections are ubiquitous and many bacteria and viruses live in the soil, water, and even on people's skin. Generally, these don't cause problems unless something changes. For example, Tetanus, caused by a bacterium called *Clostridium tetani* living in the soil, occurs when the bacteria get into a cut in the skin of an unimmunized person [15]. Infection results in muscle spasms including lockjaw. Records of tetanus infections date back to the 5th century BCE, but Tetanus infections are uncommon now primarily because vaccines are available to prevent them. Even before the vaccine, however, only a few people got tetanus. Because tetanus infections come from contact with soil where the bacterium lives, and not contact between people, it does not have the ability to cause an outbreak. On the other hand, unlike Tetanus, outbreaks of Varicella zoster or chicken pox occur. Varicella is endemic in the United States as there continue to be a small number of cases occurring regularly [16]. Varicella spreads from person to person and therefore, the possibility of outbreaks exists. Most populations have immunity to Varicella either through vaccination or immunity after contracting the disease. Before the vaccination was available in the 1990's, many people acquired immunity through infection during childhood. While outbreaks still occur especially in populations with low vaccination rates, these generally remain local in nature and do not spread to a larger geographic area.

Influenza or flu, however, has the ability to spread throughout a geographic area and create outbreaks, epidemics and pandemics [17]. Natural immunity to flu develops from infection, however, it is not effective in preventing severe disease, especially during flu seasons as the flu virus mutates or changes throughout the flu season as well as into the next season. Vaccines to protect the population from severe disease must be crafted annually to match the current expected virus variants. Small outbreaks or sudden increases in the number of cases may occur generally in a community or geographic area especially when only a small proportion of the population has been vaccinated against flu. Flu can also create more widespread epidemics where many people are infected in a short period of time, such as occurred in 1918, 1957 and 1968. Other viruses can produce epidemics, such as the original coronavirus labeled Severe Acute Respiratory Syndrome (SARS) which killed close to 800 people in 2003 [18]. Both influenza viruses and coronaviruses have caused global pandemics when the infectious agent is novel or new and there is little to no immunity in the population [17,18]. The lack of immunity and the agent's virulence allow it to spread throughout the population infecting many people and resulting in large scale illness and death, leading to social and economic disruption.

The 1918 flu pandemic lasted from February, 1918 until April, 1920, infected about one-third of the world's population and killed about 50 million people, or 2.7% of the population worldwide [10,17]. This flu pandemic preferentially impacted people aged 25 to 40 years as well as the very young and those over 65 years old. At that time there was no flu vaccine available and there was only a rudimentary understanding of the mechanism of virus and disease spread, although masks were used to attempt to stem the spread. Now, the availability and use of the annual flu vaccine, designed to address the changing flu virus, likely has prevented a flu pandemic recurrence since the last major one in 1968. Novel strains of other viruses appear periodically, often originally from an animal source, and sometimes cause human infection. While the 2003 SARS virus only generated a limited epidemic, in December, 2019 another coronavirus, severe acute respiratory syndrome coronavirus 2, also known as SARS-CoV-2 or COVID-19 emerged, resulting in a global pandemic [10,11]. It has particularly affected people more vulnerable to infections, including older adults, people with certain chronic conditions such as obesity and diabetes, and people with disabilities. In addition to the human costs, COVID-19 impacted the world economically and socially, with many businesses shuttering for a period of time or for good, and quarantine and sheltering at home resulting in increased social isolation.

As of July 2021, COVID-19 has caused 185 million cases worldwide and four million deaths, representing 2.3% and 0.05% of the population, respectively [11]. This coronavirus has modified over time, creating multiple variants with differing levels of infectivity, morbidity and mortality. Global lockdowns with social distancing, requirements to wear masks, and limitations on gatherings slowed the virus where there was adherence. Rapid development of effective vaccines provided immunity for those that got the vaccine and further slowed the virus in locations with a high number of vaccinated people. This creates a phenomenon called "herd immunity" where a large part of the community is immune to the disease limiting spread from person to person. This concept is the basis of many vaccines including Varicella, flu and measles [8,16,17]. With COVID-19, vaccine hesitancy has diminished the vaccine's potential impact, resulting in higher numbers of cases, mortality and long term consequences from the virus. Nonetheless, the impact of the virus on the world population has thus far fallen short of the impact of the 1918 flu pandemic, likely related to the implementation of multiple public health disease control strategies. It is thought that the COVID virus, like the SARS virus, originally came from an animal such as a bat or pangolin and jumped to infecting humans. As this occurs regularly, there exists the possibility of the similar occurrence of other epidemics or pandemics, making the role of public health in the management of these types of events crucial.

Natural disasters

Most of the disasters discussed so far represent infectious diseases. Weather and climate-related and geological events are classified as natural disasters. Hurricanes in the North Atlantic, cyclones in the North Pacific, and earthquakes across the globe result from natural processes of the earth. These events affect areas of the world based on their geologic structures and tend to recur. Hurricanes and cyclones generally impact coastal regions with high winds and torrential rains resulting in flooding and wind damage. These events are seasonal and in some seasons areas may be affected by multiple events in a season. Meterologic studies and practices have made the tracking of these events more accurate often allowing earlier warnings and evacuation with particularly powerful storms. Earthquakes also tend to recur in particular geographies. These areas often have geographic structures called fault lines with breaks between two pieces of rock that slip and cause earthquakes. Earthquakes result in damage to buildings and infrastructure such as roads, bridges, water supply and power lines. Earthquakes occurring under the ocean can trigger tsunamis with a series of huge waves that devastates coastal areas. Changes in regional or global climates can increase the frequency and intensity of these natural events, resulting in more property damage and loss of life. Some of these changes have been caused by humans over time through the use of fossil fuels, but the results lead to changes in natural processes.

Human-caused disasters

Not all disasters are natural and many kinds of disasters are human-caused. Bombs, shootings, terrorism including cyber threats, and nuclear meltdowns constitute some examples of human-caused disasters. A number of examples of bombings have occurred in the United States, including the Boston Marathon bombing (2013), the New York City World Trade Center bombing (1993), which did not destroy the building, but caused death, injury and significant damage, and the Oklahoma City bombing (1995) that destroyed a federal building and resulted in multiple deaths and damage to the surrounding area [3,19,20]. Less than ten years later on September 11, 2001 another human-caused disaster referred to as "9/11" destroyed the twin towers of the New York City World Trade Center buildings [21]. This event, an act of terrorism, occurred when two commercial planes were taken over by terrorists and flown into the towers, razing both of them and resulting not only in the death of about 3000 people, but also the long term disability of many first responders and victims due to exposure to the destruction of the buildings.

Local community emergency events also may stem from human-caused actions. School and other community shootings such as those at Columbine High School in Colorado, Sandy Hook Elementary School in Connecticut, and a movie theater shooting in Colorado represent local disasters that ripple through the country. In the first few months of the COVID-19 pandemic, school shootings increased more than in other years worrying parents and officials about what might occur once children returned to school [22]. While these are local events, national news coverage about the circumstances, including the number of people injured or killed, expands the event's impact across the country. Other smaller events such as fires, however, may have a smaller circle of impact geographically and in terms of the number of affected people, but nonetheless, require advance planning for response and recovery.

Cyberattacks

Attacks on technology represent another category of emergency, with very different impacts. Cyberattacks can be defined as unauthorized access of the computer network system of an entity, such as a business or government, with the intent to damage or disrupt digital operations and data, steal data for ransom or to make a political statement [23]. The motives behind cybercrime include activism, theft, spying or sometimes just a prank. In 2020, the five most common mechanisms used in cyber threats and crimes were social engineering including phishing, ransomware, distributed denial-of-service (DDoS) attacks, third party software, and cloud computing vulnerabilities [24]. Some cybercriminals remotely place software or malware on a computer or into a system to disable it via viruses, Trojan horses, and worm methods. Others, such as distributed denial-of-service disrupt service by making the host server unavailable to the users. Over time, remote work and cloud-based information technology (IT) have become more widespread. This protects businesses from IT service disruption with the local physical impacts of a natural disaster such as an earthquake. However, as digital operations become more sophisticated and dispersed in the cloud and work is more commonly conducted remotely, the risks of being hacked increase, making cyber intelligence and risk management a critical business function. IT planning constitutes an important consideration for emergency management and business continuity in the event of a disaster.

A myriad of types of disasters can occur with various characteristics including natural or human-caused, local or widespread, long or short, etc. In considering disasters, however, they all have some general similarities that planning must address. Capitalizing on this, disasters began to be approached as "all hazards" rather than by individual type [25]. This tactic creates a framework for planning for and responding to any type of disaster with a common communication and leadership structure, coordination of resources, and a single plan to follow. This strategy also streamlines the planning process as developing an individual plan for each different disaster type requires substantially more effort and time than a single plan. However, this organization of emergency planning and response emerged from a number of iterations over time. Studying the history of emergency response places the current approach to emergency response in context.

History and evolution of emergency management

Emergencies, disasters and responses to them have existed across time. However, until recently, no organized approach existed. In the United States, the earliest involvement of government in local disasters occurred in 1803 when Congress passed an act to provide monetary aid to a New Hampshire town destroyed by fire [26]. The next significant federal government involvement in local and regional disasters and disaster prevention didn't occur until the 1930s when government loans became available to repair public facilities after disasters. About the same time, the Tennessee Valley Authority (TVA) was created to build dams to produce hydroelectric power and reduce flooding in the Tennessee Valley region. A few years later, World War II brought with it the fear of a US mainland invasion which led to the rise of local civil defense [27]. Civil defense activities included home survival training and blackout curtain patrols by resident volunteers, but were often also linked to local police or fire departments. In order to harness the roughly 1000 local civil defense councils made up of around 10 million volunteers from 44 states, in 1941, President Franklin D. Roosevelt launched the Office of Civil Defense which housed its Civil Defense Corps. The goal of the Corps was to coordinate activities and partnerships between volunteers and local municipal governments in order to perform specific civil monitoring and homeland defense tasks during war, including shelter instruction, camouflage for vital facilities, and plans for evacuation readiness [28].

The advent of the Cold War in the wake of World War II brought a new, more organized approach to emergency management, primarily directed at the perceived threat of the time, namely warfare. The Cold War defined a time period where a state of political enmity existed between the Soviet-led Eastern or Communist Bloc and the Western Bloc represented by the United States and Western Europe. This period from the end of World War II to the early 1990s was characterized by threats, media propaganda, a nuclear arms race and the funding of proxy wars in other parts of the world such as the Vietnam War in the 1960s and the Soviet occupation of Afghanistan beginning in the early 1970s. During this time and in response to the potential for nuclear war and related disaster, the presence of local civil defense programs expanded and grew in communities and states across the country [27]. The US federal government provided technical assistance to local civil defense programs through the minimally staffed Federal Civil Defense Administration (FCDA). In addition, a separate federal agency, the Office of Defense Mobilization (ODM), was tasked with stockpiling and producing needed goods and materials to prepare for potential war. Emergency planning was one of the tasks assigned to the ODM. In 1958, these two entities merged to become the Office of Civil and Defense Mobilization which was housed in the Department of Defense (DOD). This series of events resulted in the Civil Defense model of emergency management. This model provides for the integration of national and local emergency planning where the federal government sets the national strategy and plan, and local and state government civil defense directors are responsible for aligning local plans with the national one and for plan implementation during an emergency [25,26].

The focus of emergency preparedness remained on the threat of nuclear warfare throughout the 1950s, a time during which there were only three major hurricanes. The 1960s brought significant increases in the number of natural disasters which caused substantial damage with economic and human impacts. In 1961, three major disasters occurred, an earthquake measuring 7.3 on the Richter scale in the Hebgen Lake area in Montana and two hurricanes, Carla on the Texas coast and Donna on the west coast of Florida. In response, President Kennedy created the White House Office of Emergency Preparedness to address natural disasters [25,26]. Additional natural disasters that decade included an earthquake in Alaska with tsunamis along the coast of California, a storm on Ash Wednesday that destroyed part of the East Coast shoreline, and two devastating hurricanes on the Gulf coast. These events led to the identification of a need for additional economic protections in regions at risk for these storms. Thus, the National Flood Insurance Program (NFIP) was created to make flood insurance available to residents of flood-prone areas. Despite the increased importance of the impacts of natural disasters and ongoing civil defense concerns, US emergency management functions and administration remained split between civil defense programs housed within the DOD and the White House natural disasters program.

The presence of emergency management and planning for various types of events increased in the federal government in the 1970s [25,26]. More than 100 federal agencies held responsibilities for emergency management for various areas. For example, in addition to the DOD's nuclear defense and flood control activities, programs for weather disasters, continuity of government and federal stockpiling and planning, power plants and imports investigations existed in the Departments of Commerce, General Services, Treasury, and the Nuclear Regulatory Commission, respectively. This fragmentation led to disputes over the Offices' jurisdictions and manifests itself in disorganization and confusion during disaster response. The 1979 Three Mile Island Nuclear Plant accident near Harrisburg, Pennsylvania brought this issue to national attention when the lack of planning and coordination became evident. The accident involved the nuclear plant's cooling system and resulted in radiation exposure to the local area related to some extent to errors in recognition of and response to the event [28]. Partly in response, President Jimmy Carter consolidated the functions for disaster management into a single agency called the Federal Emergency Management Agency (FEMA). The FEMA director reported directly to the President, giving the agency significant visibility within the federal government. FEMA remains the primary federal agency responsible for "leading the Nation's efforts to prepare for, protect and mitigate against, respond to, and recover from the impacts of natural disasters and man-made incidents or terrorist events" [29]. In all of these iterations of emergency management structure and practice, the conversation about the impact of disasters on people with disabilities did not arise.

Emergency planning models

Emergency planning and response arose during and after World War II out of civil defense and the need to protect the population against potential warfare. The experiences with fire bombs during the war and the possibility of nuclear bombs drove the preparedness and response planning. Three fundamental practices were intended to minimize a nuclear bomb impact, time, distance and shielding, which require getting people as far away from the blast site as quickly as possible and shielding them from the blast and radiation exposure by using lead bunkers or natural blocks like mountains [27]. In the civil defense model, law enforcement or fire departments generally lead the efforts by employing the key approach of evacuation and relocation. Consequently, planning incorporated the elements necessary for that purpose, roadblocks to facilitate one-way traffic, tow trucks, refueling stations and a law enforcement presence. During President Dwight D. Eisenhower's administration (1953-1961), bunkers with air lock systems (reinforcement against the potential force of an explosion and fallout) were constructed and became Emergency Operations Centers (EOCs). In order to ensure continuity of government, all operations were to be moved to these EOCs in the event of an attack. EOCs contained attack warning systems, space for volunteer amateur radio operators to assure a second form of communication with the ability to dispatch orders from inside. The EOCs were intended to support the staff for a few weeks during and after an attack to ensure government continuity. Weather stations were also included as tracking the wind, humidity and precipitation was important to predict the impact of a bomb and dissolution of fallout.

Public administration model of emergency management

As the threat of nuclear war dimmed and the occurrences of natural disasters rose, the nature of need for emergency response changed. Emergency management functions evolved to include more public education for disaster preparedness. In places like San Jose, California the civilian staff became part of the local fire department where education about fire risks, drills and evacuation planning were part of the standard functions [27,30]. Over time these functions became part of the municipal government, creating the Public Administration model of emergency management. Selves [31] notes that the Public Administration model envisions emergency management as a separate function of government administration rather than part of an emergency services agency such as the fire department. Its function is to assure continued government operations during a crisis. As well, Selves also describes this role as providing "a proactive framework and pre-planned mechanisms which allow government to operate in an effective and integrated manner during a crisis" [31]. This model views emergency management as a discipline with an academic and professional focus on research and debate. In this model, various stakeholders, including politicians and the media, play roles that require study to understand the influence on emergency events. In addition, the model places the coordination and integration of all public and private emergency management efforts under a single framework.

Emergency Services model

Another model, called the Emergency Services model, focuses on an emergency services agency's conduct, management and coordination of disaster response [26]. In practice, it involves the role of emergency services, largely centered on law enforcement. Important response functions include moving people and equipment during an event with drills to prepare for that, but not coordination between departments. Politicians and media do not have a role and are often seen as being in the way. Members of the community are expected to play the role of disaster victims, but not an active role in planning. Instead, interactions are between the entities providing the emergency services. Emergency management often is an added duty within the emergency services agency rather than a primary function, making it less important than the agency's primary purpose. For example, law enforcement's primary responsibility is the enforcement of laws and maintaining public order. Its main role in managing public safety is protecting the public from criminal activity. In the Emergency Services model, emergency managers may be relegated to a more passive role as a resource rather than having active involvement during a disaster, with the "lights and sirens" taking precedence. With this emphasis, issues such as mitigation, public awareness, continuity of government, and other matters perceived as non-emergent receive less attention. The Emergency Services model stresses the importance of the actual response, with less weight placed on planning, educating, or mitigating the impact of the disaster through actions with longer term effects.

Other models of emergency management

Additional models conceptualizing emergency management exist, and these fall under a few different categories based on the model's goals related to disaster management, risk assessment and management, or understanding underlying causes. Models such as the Pressure and Release (PAR) model, developed by Blaikie and colleagues in 1994, seek to explain the relationships and underlying root causes of hazards and vulnerability [32–34]. Linear risk management models, such as that used by the

Canadian Standards Association, employ the strategy of identifying and analyzing risk to determine if that risk warrants mitigation [35]. This approach can be adapted to disaster management. Other entities such as CARE, an international, non-governmental organization (NGO) that combats poverty and hunger and provides emergency, humanitarian relief across the world, approaches disaster response as one of the many "shocks and stressors" that impact the economic well-being and security of the family, household and the community [36]. The Comprehensive Emergency Management model, which appeared in the 1980s, concentrates on the disaster management and the disaster cycle [34]. These examples will be discussed in more detail below, and are only a few of the myriad conceptual models developed for thinking about disaster management as a field.

The Pressure and Release (PAR) model

The Pressure and Release (PAR) model, developed by Blaikie and colleagues, is one of the most well-known and accepted models for explaining the role of disaster social risk and vulnerability [32,33]. Over time variations in the model's conceptualization have appeared, but it remains a stalwart. The model and its variations explain how disasters are shaped by pre-disaster social, economic and cultural structures and processes in a community and society. The model focuses on cause and effect to explain the role that those pre-disaster factors have in forming vulnerabilities and how the intersection of hazards and vulnerability leads to disaster. These models attribute the magnitude of the disaster's effect on vulnerable populations to vulnerability pre-disposition based on socioeconomic and political policies and practices. The PAR model examines risk as a function of hazard and vulnerability where vulnerability is measured by exposure and susceptibility.

Examining the model more closely, Fig. 1 depicts the two opposing forces: hazard as represented by a physical exposure to a disaster against three levels of vulnerability [34,35]. The three levels of vulnerability, root causes, dynamic pressures, and unsafe conditions characterize the progression of increased vulnerability from causes rooted in lack of access and resources to the creation of unsafe conditions. The model posits that pressure can come from either side of the equation, hazard or vulnerability, but the only way to release it is to decrease vulnerability, often accomplished by increasing resilience. Morris et al. [37] notes that economic and political systems form the root causes leading to social inequities and vulnerability within a population based on gender, race, and disability. These inequities produce unequal access to resources and opportunities as well as unequal exposure to hazards. The impact Hurricane Maria (2017), a Category 5 storm, had on Puerto Rico is illustrative. Studying the death rates after Maria, Morris and colleagues identified that the death rate caused directly by the storm was relatively low. However, in the aftermath of the disaster, deaths due to natural causes spiked. Applying the PAR model to this event, Puerto Rico has a long history of environmental degradation, unregulated construction, improper disposal of toxic wastes, such as mercury from military contamination, and lack of infrastructure including reliable electricity, clean running water and other resources. These elements mostly affected the population with limited financial resources that lived in the areas where the construction and waste disposal occurred. This exposed that population to health dangers such as pollution, infectious diseases and poor nutrition, which in turn harmed their health because of cumulative exposure to environmental toxins. This kind of exposure is associated with higher rates of heart disease, diabetes, asthma, and high blood pressure as well as earlier development of these conditions and a shorter lifespan [38]. In addition, Puerto Rico suffers from high rates of poverty and lacks the safety nets such as Social Security Income (SSI) available in other parts of the United States. It also

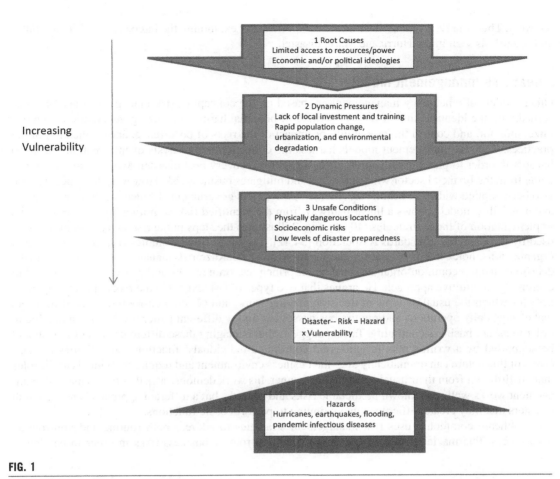

FIG. 1

The Pressure and Release (PAR) model.

Adapted from Wisner Canadian Standards Association. Risk management: guideline for decision makers; 2002. http://lib.riskre-
ductionafrica.org/bitstream/handle/123456789/743/risk%20management.%20guidelines%20for%20decision-makers.%20a%20
national%20standard%20of%20canada.pdf?sequence=1 [Accessed 30 September 2021].

has a lower federal match for the Medicaid program and a cap on federal funds it receives, requiring the territory to fill the gap with its own funds. In addition, the adult poverty rate in Puerto Rico is twice that of the poorest US state with the poverty rate of adults with disabilities in Puerto Rico being 1.85 times the US average rate [39]. As well, examination of the causes and effects of hazards and vulnerabilities created by the Hurricane Maria disaster reveals that many of these have the same underlying origin [33]. Unregulated development of Puerto Rico's coasts with improper toxic waste containment not only impacted population health, but also disrupted the natural coastal ecosystems that protect the island from storms [38]. The conflagration of vulnerability related to poor health, lack of resources, and poverty in the face of a Category 5 hurricane created a significant disaster that will takes years for recovery. Thus, the PAR model illustrates the importance of examining the intersection of vulnerability and hazards as seen in the Hurricane Maria disaster.

Linear risk management model

Other models of emergency management are based on the concept of risk management and are concerned with the identification of risks or sources of potential harm, strategizing interventions to minimize, monitor, and control the probability or impact of the risks of occurrence, and application of the practices. The risk management models harness the collection and analysis of information related to hazards in order to plan the utilization of resources for recovery post disaster. Much of this work has come from the financial sector which analyzes and mitigates business risk. However, the concepts have also been applied widely in health care, human services and government. Termed the "linear risk management," this model follows a linear pathway from the identified risk, to mitigation strategies, to the implementation of those strategies [40,41]. Fig. 2 illustrates the steps in the risk management pathway used by the Canadian Standards Association [34,42]. Tarrant [40], who studies risk management and organizations, notes that it is vital for organizations to standardize risk management in order to make decisions using a common organizational set of principles, frameworks and processes, as opposed to exercising an intuitive approach. He argues that two types of risk exist: (1) the risks of everyday business for which the usual patterns of decision making work, and (2) the non-routine risks of disasters that change daily operations and require businesses to employ different patterns of decision making in order to assure business continuity. Tarrant suggests that managing these different types of risk should be supported by a written plan to guide and coordinate individuals' function and activities because crises of this nature can dramatically alter the business environment and require individuals to fill roles that are different from their usual ones. In this manner, his work demonstrates that traditional risk management works well with known predictable risks and hazards, but a different approach is needed with disasters that may create difficulties in continuing to perform basic functions.

Healthcare commonly uses risk management strategies to address both routine and non-routine occurrences. Pharmacies plan for both routine and non-routine business risks in order to be able to

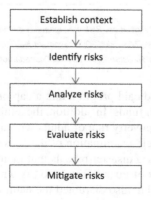

FIG. 2

The Linear Risk Model [42].

provide medications for those that need them. Using the example of a common medication to treat seizures, carbamazepine, can illustrate how these two strategies might differ. In current pharmacy practice, because of space constraints and other considerations, pharmacies stock only the amount of a medication needed for typical use. They track the volume of each medication dispensed to determine how much they need to keep on hand. Most pharmacies use "just in time" ordering where they place regular orders for commonly used medications for daily or almost daily deliveries. Suppose a delivery truck with the daily order for carbamazepine has an accident and its shipment is destroyed so it won't reach the pharmacy. This represents a common risk in this business as local truck delivery is common and traffic accidents regularly occur. Without the seizure medication from that delivery, the pharmacy may run short of carbamazepine. Therefore, the pharmacy needs a plan to address days where demand exceeds the available supply. One mitigation strategy could include giving patients some of the pills for their prescriptions and filling the remainder when the next shipment arrives. This way the people with seizure disorders would not run out of seizure control medication and the pharmacy would be able to provide everyone with enough medication to last until the next shipment arrives. As an alternative, the pharmacy could have a back-up source for quick filling of small orders when they run short. Regardless of the solution, this is a frequent and routine risk for which mitigation doesn't significantly change business operations.

If instead, a fire destroys the manufacturing plant that makes carbamazepine, then all of the existing medication in production in that plant is destroyed and the ability to make the medication halts until the plant can be rebuilt or repaired or the manufacturing process is moved to another site. This will create a shortage of carbamazepine for a significant period of time not just for this pharmacy, but for all of the pharmacies that obtain their stock from this plant. This is a disaster for the plant as well as the pharmacy and is not a routine business risk. For the pharmacy, the issue is how to procure carbamazepine for their patients to prevent seizures. The solutions from the previous example, such as waiting until the next order arrives, won't work here because the carbamazepine source has been destroyed and no more medication will be made for a period of time. Rather, this requires planning for alternate sources for procuring the medication. The pharmacy could work with its distributor to obtain the medication from an alternate plant or company or to use a different brand in lieu of the destroyed product. If the problem becomes more extensive and sufficient amounts of the medication from other plants or similar drugs are not available to meet the demand, then the federal Food and Drug Administration (FDA) may step in to help effect a solution [43]. This approach parallels the tactic used in emergency management in the United States with the federal government stepping in when first local, and then state solutions are no longer effective either because of a lack of resources or an incident too widespread to be solved by the state and locally.

The Risk Management model is used frequently and focuses on planning primarily to address known daily risks with less attention paid to unusual, catastrophic risks. In contrast, another commonly used model, Comprehensive Emergency Management (CEM) focuses on planning for and managing disasters. This model is the foundation for the current public health framework. Defined by McLoughlin (1985) [44], the model evolved from previous iterations focused on preparedness and response to better meet the needs of the post-Cold War era. The new model adds mitigation and recovery to preparedness and response to form the four pillars of the framework [34,44,45]. In 1993 FEMA director James Lee Witt shifted the emphasis from preparedness and recovery to mitigation [25]. Prior to this, in 1979 at the juncture of FEMA's formation, the National Governor's Association developed state guidance noting that preparedness, response, mitigation and long term recovery are closely linked as well as being important emergency management elements [45]. These four elements remain the pillars of many emergency management models, although other models incorporate different elements. A number of authors have described and illustrated the relationships between these elements in variations of this model developed over time. Examples of these variations include models by the Tuscaloosa County, Alabama Emergency Management Agency, the Manitoba, Canada Health Disaster Management, and the United Kingdom Integrated Emergency Management [46–48]. Some of these models primarily use the four elements for emergency management and illustrate how they relate one to another. Others expand on these elements and include additional important emergency management activities from other models, such as the Risk Management model previously discussed. These different models provide some diversity of thought and focus, but with similar themes.

The Tuscaloosa County Emergency Management Agency (TCEMA) model

The model used by the Tuscaloosa County Emergency Management Agency (TCEMA) shows the classic picture of the relationship between the original four elements: preparedness, response, mitigation and recovery to which they have added a fifth element, prevention [46]. This model, an example of the Comprehensive Emergency Management (CEM) model, is illustrated in Fig. 3. Tuscaloosa County, Alabama provides disaster training and education, guidance for personal and household disaster plans and coordinates disaster response. Starting with the response stage, this model illustrates a circular, sequential path through the five elements and back to the response stage in a continual pattern. Sometimes termed a logical model, it defines the four elements and the role of each in disaster response, emphasizing the events and activities that occur [49]. The current version of the TCEMA model incorporates a fifth element: prevention. The TCEMA advocates for first responders to know and understand the actions that need to happen on the ground in a disaster. They provide coordination, communications, resources, and information to inform all stakeholders including elected officials of

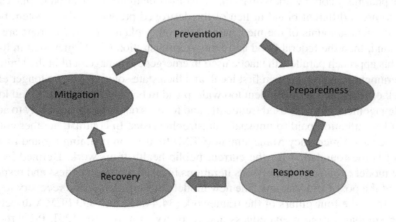

FIG. 3

Tuscaloosa County EMA.

*Adapted from Tuscaloosa EMA. Tuscaloosa county emergency management cycle; 2003. https://www.tuscaloosacountyema.org/about/
evolution-of-emergency-management [Accessed 30 September 2021].*

progress and to share ongoing situational awareness. The TCEMA acts as an intermediary between the county and state and federal emergency management when they are involved. After the disaster the TCEMA also assists with recovery.

The Manitoba model

In Canada, the Manitoba model used by the Manitoba Health Disaster Management branch of the Manitoba public health department focuses on health sector disasters [47]. This integrated model represents a cyclical process incorporating 6 elements including hazard assessment, risk management, mitigation, and preparedness, and incorporates a strategic approach plus quality improvement. The cyclical model begins with a strategic plan defining the functions and responsibilities for each of the other elements including in a quality improvement structure to monitor, evaluate and improve the program by learning from what went right and what went wrong. Fig. 4 shows a visual depiction of the

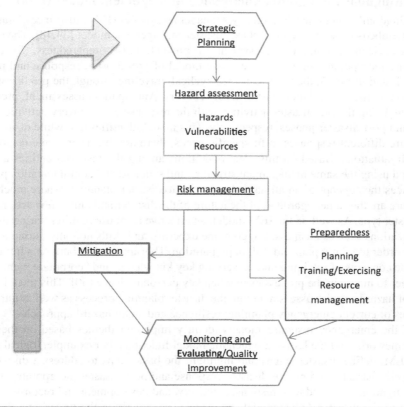

FIG. 4

Manitoba, Canada Health Disaster Management.

Adapted from Manitoba Health. Disaster management model for the health sector, guideline for program development; 2000. https://vdocuments.net/disaster-management-model-for-the-health-manitoba-health-disaster-management.html [Accessed 30 September 2021].

Manitoba model. The purpose of the hazard assessment is to identify the threats, vulnerabilities, and resources to the population. It incorporates the intersection between community vulnerability including social determinants of health or social risk factors, community coping abilities and the presence of extreme events. Risk management determines the possible impacts and develops potential intervention options. The mitigation element attempts to reduce the degree of the impact that hazards could have on the population and the preparedness element incorporates planning, training, drills, and resource management. After each drill or disaster, continuous quality improvement requires performance evaluation regarding what worked and what didn't. The model focuses on the balance between flexibility and preparedness to strengthen the ability to adapt pre-planned activities to meet the uncertainties of a disaster. This model recognizes that the health sector has a partner role in community disaster management and that disaster management needs to be integrated into the broader health program mission to "protect, preserve and promote the health of all" [47].

The United Kingdom's Integrated Emergency Management model (IEM)

The United Kingdom's Integrated Emergency Management model (IEM) also incorporates six areas of activity, but embeds these into a three phase disaster management model [34,48]. The six areas, (1) anticipation or horizon scanning, (2) assessment, (3) prevention, (4) preparedness, (5) response, and (6) recovery, are incorporated into the three phase model of pre-disaster, response and post-disaster. In this model, found in Fig. 5, the process is also cyclical, traveling through the pre-disaster period to the response to the disaster and then post disaster recovery. Anticipation, assessment, prevention, and preparedness make up the pre-disaster activities with the response and recovery activities comprising the recovery and post-disaster phases, respectively. This model identifies that while disaster situations vary and require different responses to fit specific needs, there are, however, some response features common to all situations. Based on this idea, operations among the involved entities are combined and coordinated using the same management structure and standardized communication patterns. This model introduces the concept of an all hazard approach to disaster planning, where much of the plan and the structure are the same regardless of the nature of the disaster and only a few actions change for a specific disaster type. As well, in the IEM model, an effective response requires partnering with community organizations such as businesses to combine expertise and skills not only during a disaster, but prior to one in order to jointly plan and drill in preparation. This approach is built on a framework of resilience and the concepts of retaining and preserving key knowledge and "corporate memory" through training in order to maintain the process even when key personnel leave [50]. This model incorporates the concepts of hazard and risk assessment into the disaster planning process as well as introducing two other principles of current emergency planning: resilience and an all hazard approach.

Studies of the emergency response models identify important themes based on their respective conceptual frameworks and the basis on which the model functions. For example, logical models such as the UK's IEM, define disaster stages and highlight the basic steps to address a disaster. Here the three phase cycle characterized by pre-disaster, response and post-disaster incorporates the actions of prevention, mitigation, preparedness, response, recovery, and development and reconstruction into the relevant phase of the disaster. Other models base their framework on the underlying disaster causes and the political, social, and economic factors that both contribute to the occurrence of the disaster and impact the disaster's severity and impact on various population segments. The Pressure and Release model exemplifies this strategy and the intersection of hazard and vulnerability illustrates how certain

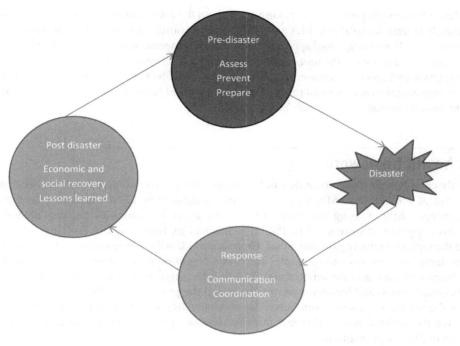

FIG. 5

Integrated Emergency Management model (IEM) from the United Kingdom.

Adapted from Essex County Fire and Rescue Service. The National Archives. Integrated emergency management and key organizations. https://webarchive.nationalarchives.gov.uk/20050112000000/http://www.ukresilience.info/contingencies/dwd/c2amanagement.htm [Accessed 30 September 2021].

populations including those with disabilities, may be disproportionately impacted by a disaster. Much of the literature concerning disability, disaster, vulnerability and resilience focuses on the concepts behind this model. Integrated models such as that used by the Manitoba public health department link disaster management and planning events and activities using strategic planning, hazard assessment, risk management, mitigation, preparedness, and quality improvement monitoring and evaluation. This type of model balances the structure of a framework with the need for flexibility and improvement in disaster management. Lastly, models exist that combine the elements of the three model types, logical, integrated and causal, to better understand and address needs throughout the disaster cycle. Cuny's model (1998) joins elements from each of these model types to paint a more complete picture of what is needed to manage successfully before, during and after a disaster [49,51].

Cuny [51,52], a well-known humanitarian and co-founder of the University of Wisconsin Disaster Management Center (DMC), was instrumental in professionalizing the field of disaster management by emphasizing the need for having trained, expert professionals in the field. His work in disaster management began with his involvement in humanitarian aid during the Biafran, now part of Nigeria, famine in 1969. Decades later, the model attributed to him and his group was published [53]. This model, informed by his years of experience participating in a wide range of disaster relief efforts across the world, offers a framework for decision-making throughout the event and into recovery in order to direct aid for both short term survival activities and longer term reconstruction and helps maintain continuity of operations [51]. His principles of approach to disaster management emphasizes practical, low cost, and local solutions, analysis of the root cause of the disaster and learning from the event through monitoring, evaluation and process improvement. Through the DMC he helped develop and deliver training and educational events to achieve that [52]. Cuny's concepts and framework provide the underpinnings for current disaster management models and practice.

Public health framework

Historically emergency management as a field emerged from civil defense, volunteer disaster relief, and response of fire and law enforcement. The field as defined by FEMA became "the managerial function charged with creating the framework within which communities reduce vulnerability to hazards and cope with disasters" [54]. Public health had its beginnings in antiquity when around 500 BCE the Greeks including the physician Hippocrates identified an association between the presence of malaria and swamps and illustrates the first recognition of a systematic, causal relationship between human disease and the environment [55,56]. During the Middle Ages, infectious diseases such as bubonic plague and leprosy were prevalent and the practice of isolation developed as a way to prevent the spread of disease even before an understanding of how disease was actually spread. As the scientific understanding of communicable diseases grew, public health adopted additional approaches to disease prevention.

Today, public health plays a role in many aspects of health promotion and disease prevention including food safety, vaccination programs, potable water supply, and safe housing, continuing to correlate factors with disease and the development of prevention strategies and treatments. In addition to infectious diseases, public health focuses on addressing risk factors that can result in chronic medical conditions. These risk factors include tobacco use, obesity, and lead exposure, which have been associated with chronic lung disease, diabetes, heart disease and other conditions. Public health creates prevention and treatment programs designed to improve population health by educating and supporting behavior change to manage or avoid the risk factors. Many of these chronic conditions lead to disability and contribute to the development of challenges with function and the ability to access the community. Populations with functional and access challenges constitute a portion of the population of people with disabilities that need to be considered during emergency planning and response.

In the past, much of emergency management occurred within the purview of public safety and rarely intersected with public health, regardless of the nature of the hazard. However, public health has gained a more active role in emergency management as the impact of disasters affecting population health and creating disparities become more prevalent. As more events that involve infectious agents, such as severe acute respiratory syndrome coronavirus 2 (COVID-19) or Ebola lead to regional or worldwide epidemics, a new domain within emergency management, public health emergency management, is emerging [57]. This new field recognizes that disasters other than infectious diseases also threaten health. When Hurricane Maria tore through Puerto Rico in 2017, it destroyed human lives as well as the housing stock, infrastructure and island vegetation [37]. The death toll attributed to the storm itself was low, about 55 people. Yet, after the storm, the rate of deaths from natural causes increased dramatically with much of that burden affecting populations of people living in poverty with chronic conditions and disability. The intersection of the tenets of public health to prevent disease and promote wellness in the population and the impacts of health-related and non-health-related disasters on population health emphasizes the importance of the role that public health systems play in the emergency management of disasters [57]. Strengthening public health systems to better protect communities and populations from disasters with involvement throughout the disaster cycle is critical for prevention as well as recovery. Using the World Health Organization's (WHO) definition of health, "a state of complete physical, mental, and social wellbeing," public health clearly contributes to the management of disasters throughout the disaster cycle [58].

Public health emergency management adds tools to emergency management to address factors related to disasters that impact health. Gathering air and soil samples, cultures or tests for specific infectious diseases, and other data provides public health with information to follow patterns of infections and intervene when and where needed. Geographic mapping of positive COVID-19 tests provided valuable information about where the infection was most active and demonstrated the effectiveness of vaccines [59]. The testing also identified new virus variants which were tracked to determine if they differed from the original virus in infectivity or seriousness of infection that would warrant changing mitigation recommendations. Public health lead poisoning programs use lead pediatric paint poisoning cases to geomap the presence of lead paint in houses by zip code. The allocation of funding and resources to remove the lead from houses in those zip codes would eliminate a source of lead poisoning in children's environments.

Other measures that public health brings to the table are medical countermeasures. Medical countermeasures (MCM) represent a variety of products that assist in the identification of, prevent the occurrence of, or treat conditions that result from a disaster [60]. During the COVID pandemic, MCM such as personal protective equipment, masks, gowns, gloves and other protective equipment were used by hospital personnel to protect themselves from infection. Vaccines, antibiotics and antiviral medications constitute another type of MCM as do other drugs used to treat or prevent disease. Potassium iodide and Prussian blue are two substances used to treat radiation exposure by binding to radioactive molecules to protect the thyroid. Public health uses MCMs and the data analysis to target communication, guidance, prevention and treatment in the event of a disaster with health implications.

As public health plays an important role not only in acute health issues, but also in chronic conditions, it makes sense to examine disasters and the inclusion and integration of people with disabilities, chronic illness and mental health conditions, and their needs related to disasters through a public health lens. Public health emergency management frameworks parallel those of other emergency management entities, for example by using the emergency management life cycle phases [25]. Focusing on the public health framework incorporates public health work to understand the epidemiology of chronic physical and mental health conditions and their intersection with disability, and brings an element of detail that is critical to know when planning with human populations. The public health framework also assists with understanding the vulnerabilities and potentials for disaster resilience among these populations. The descriptive term "access and functional needs" (AFN) has been adopted to represent this diverse group of people in an attempt to describe and address their needs. However, it is important to understand who people with access and functional needs are and where they live to be able to engage them in advance planning for disasters to avoid repeating past occurrences. Thus, a brief introduction to vulnerability and resilience follows with more detail in subsequent chapters.

Vulnerability and resilience

Chronic illness and disability are frequently associated with vulnerability and this has increasingly become a focus of emergency planning. Harken back to the Pressure and Release model to begin to examine the push and pull between the concepts of vulnerability and resilience, one which leads to adverse reactions to hazards and the other facilitating the ability to absorb and recover from the disruptive event. Buckler and colleagues [61] undertook a study of vulnerability and resilience to identify which populations might be vulnerable in order to incorporate a vulnerability assessment into emergency planning. They began with a simple definition of vulnerability used in Emergency Management Australia, "the susceptibility and resilience of a community and environment to a hazard" [61]. Using this definition they attempted to identify vulnerable populations and factors associated with vulnerability, but continued to find discrepancies between what was expected and what was found. For example, during a gas shortage in Australia in 1998, they expected that the elderly would be more affected by this than younger people because of their physical frailty. However, the elderly coped better with than expected as they were able to draw on their past life experiences with deprivation during the World Wars and the Great Depression. The researchers concluded that social, economic and cultural factors play a much larger role in vulnerability than individual demographic characteristics. Others have come to similar conclusions. Wisner noted that these social, economic and cultural factors generate vulnerability in populations and often define how they respond to disasters [32]. In this manner, health disparities experienced by people with disabilities also reflect inequities that go beyond genetics and health behaviors [61]. Those factors embedded in the community bring with them not only the power to create vulnerability, but also the solutions to minimizing vulnerability and building community and individual resilience.

These factors impact vulnerability as they shape the lives of people with disabilities in a number of ways and contribute to their ability to respond to a disaster. People with disabilities are more likely to live in areas with poverty and be unemployed or underemployed, limiting their ability to meet their everyday needs in terms of food and shelter [62,63]. Amassing an emergency supply of food in addition to everyday needs requires economic resources that they may not have. In addition, poorer areas are more likely to have poorly built and maintained housing situated near areas likely to be most affected in a disaster. Low levels of education as well as limited ability to access information available to the general public because of a disability, poor reading or comprehension skills, and/or lack of available accessible informational materials contribute to difficulties in participating in planning or knowing what to do during a disaster. Limited access to healthcare, social services and the lack of a safety net for these populations create risks common to people with disabilities which can be compounded by a disaster [64]. These along with narrow social networks and reduced access to political power compound existing increase vulnerability. Therefore, identifying these risks provides opportunities to fill the gaps as part of emergency planning and response and strengthen this population's resilience [65].

Disasters can disrupt social circles and supports; building social capital within the population can provide protection. While populations of people with disabilities, chronic physical and mental health conditions are seen as vulnerable populations, some of this vulnerability may be derived from the social, economic and cultural factors that Buckler discussed and not solely from the innate nature of their health condition or disability [61]. Focusing on emergency management, Smith and Jolley [62] identified four reasons that vulnerability and inequality have increased and advocate for an inclusive approach to using social capital to minimize vulnerability. These reasons include: (1) governments and relief organizations lacking knowledge and understanding of disabilities; (2) emergency managers failing to include people with disabilities in emergency management and recovery; (3) inadequate resources and preparedness including lack of accessible shelters; and (4) the perpetuation of disparity through stigma and discrimination. These themes pervade the research around people with disabilities and can be used to begin to think about barriers to resilience and recovery faced by people with disabilities. Understanding who these groups are, where they live, who they live with, how they are supported, and what their strengths are, helps not only to identify the hazards and risks that they face, but also offers the opportunity to incorporate them into all aspects of emergency management. People with access and functional needs bring their lived experiences facing social and environmental barriers on a daily basis as well as their experiences living in the local community, both of which are valuable when considering potential impacts and needs in the event of a disaster [66]. Inclusive practices can help expose the impact of social, political, economic and cultural factors on people with disabilities and begin to address the resultant vulnerability in order to build resilience.

Vulnerability and resilience often become linked and the identification of the root causes of vulnerability can lead to solutions that mitigate them promoting resilience. Emergency management focuses much of planning and response efforts on physical and environmental infrastructure as well as continuity of business operations. However, social factors contribute to the impact of a disaster on the population and deserve consideration. Martin [67] developed a framework, the Social Determinants of Vulnerability Framework, based on research in Boston, to outline the most common social factors that cause vulnerability related to disaster. The components identified as drivers include lack of access to post-incident services; displacement; injury, illness, and death; property loss or damage; domestic violence; and loss of employment. Some of these factors may be unique to Boston, however, this strategy to identify what creates vulnerability in a population helps determine the solution. Applying this approach combined with inclusive planning can create successes that minimize impact and increase community resilience. For example, in 2011 the Centre for Disability in Development (CDD) in Bangladesh studied the effects of disasters on people with disabilities which led to some significant changes in their emergency planning approach [65]. Not only did they including people with disabilities on planning committees, but their presence lead to construction of an accessible rescue boat guided by people with disabilities and building schools and homes above the flood line to protect them from displacement in future floods. Both of these projects benefit the general population as well the people with disabilities. Without the presence of people with disabilities on the committee it is unlikely that these solutions would have been considered. While vulnerable populations may have more pre-disaster stressors, inclusive practices and equitable and accessible solutions can help build individual and community resilience.

FEMA's National Preparedness Goal of "a secure and resilient nation with the capabilities required across the whole community to prevent, protect against, mitigate, respond to, and recover from the threats and hazards that pose the greatest risk" incorporates this concept of resilience as well as that of whole community [25]. The National Health Security Preparedness Index measures public health preparedness progress nationally and at the state level and identifies strengths and gaps [68]. This index uses a preparedness measure made by combining measures from multiple sources about a broad array of factors including infectious diseases and vaccination rates, aging infrastructure, antibiotic resistance, cyberattacks, and extreme weather events. The index measures the preparedness of the United States and individual states in a variety of actionable areas to track progress in planning and readiness. The measures health related measures such as health security, healthcare delivery, countermeasure management including drug and vaccine stockpiles, and food and water supply safety. Other measures gauge the adequacy of community preparedness, information and incident management. The national and state scores have generally improved over time since the Index began in 2013, although the current 2020 global score of 6.8 is significantly below the total possible score of 10 and even well below a score of 9, thought to represent adequate planning and readiness [68]. The increases in disasters and health threats in 2019 include emerging infectious diseases such as Zika, Ebola and COVID-19, extreme weather events, aging infrastructure including bridges and roads, global patterns of travel and trade, and cyber-security susceptibilities put the United States and the world at greater risk for damage to social structures, the economy and health. Significant variability exists between US states and regions, with patterns of lower health security scores in the Deep South and Mountain West regions than in the Northeast and Western Pacific. These regions also have fewer available resources to respond to a disaster because their populations are more rural and tend to have more people with low and moderate incomes. This differential effect can be seen in the population of people with disabilities as well as the general population. Addressing these issues requires not only identifying vulnerabilities, but also learning what helps build resilience. Current practices in serving people with disabilities focus on self-advocacy, self-actualization, and providing support to live as active community members. This concept of inclusion and the use of inclusive practices can be one tool to move closer to achieving whole community planning that truly represents the whole community.

Conclusion

Emergency management models serve as frameworks to organize the needed information to respond to and recover from a disaster. As well, they provide a conceptual basis for considering the root causes of poor response and recovery in order to fill the gaps and build capacity and thus improve response and recovery. Disaster and post-disaster experiences of people with access and functional needs indicates that multiple factors impact their ability to plan and respond, creating potential vulnerability. Many of these factors stem from social, economic and political divides between people with and without disabilities. While mitigating those factors in the long run will require much time and effort, recognizing them and accounting for them during the planning, response and recovery from disasters will help build resilience for individuals, but also for their communities. Inclusive planning, prevention and mitigation strategies hold the potential to minimize disparities in disaster impacts. Whole community and all hazard approaches to emergency planning, response, recovery and resilience that integrate public health, mental health and social and disability systems to support and engage communities, individuals with access and functional needs (including those with chronic illness and mental health conditions and disabilities), and their support infrastructure will lead to strong, inclusive resilient communities prepared for disasters for everyone.

Questions for thought

1. Using the example of the plant fire destroying a supply of a medication and potentially creating shortages, put yourself into the place of the drug company CEO and develop a mitigation plan for this type of event.
2. Choose a type of disaster and apply the Pressure and Release (PAR) model to it.
3. Take the same disaster from Question # 2 and apply one of the other models presented to it.

References

[1] Center for Excellence Disaster Management and Humanitarian Assistance. Disaster management overview definitions, Training DHMA 101. https://www.cfe-dmha.org/Training/DMHA101/Disaster-Management-Overview-Definitions [Accessed 30 September 2021].

[2] The International Federation of Red Cross and Red Crescent Societies (IFRC). What is a disaster? 2021, https://www.ifrc.org/en/what-we-do/disaster-management/about-disasters/what-is-a-disaster/. [Accessed 30 September 2021].

[3] Ford B, Smith GB, McShane L. Police narrow in on two suspects in Boston Marathon bombings. The New York Daily News 2013;(18 April). https://www.nydailynews.com/news/national/injury-toll-rises-marathon-massacre-article-1.1319080. [Accessed 30 September 2021].

[4] The University of Utah. Earthquake FAQs; 2021, https://quake.utah.edu/earthquake-information-products/earthquake-faq#:~:text=Generally%2C%20only%20seconds.,intermittently%20for%20weeks%20or%20months. [Accessed 30 September 2021].

[5] The University of Rhode Island. Hurricanes: science and society; 2020, http://hurricanescience.org/science/science/hurricanelifecycle/. [Accessed 30 September 2021].

[6] National Weather Service, National Oceanic and Atmospheric Administration. What is a nor'easter? https://www.weather.gov/safety/winter-noreaster. [Accessed 30 September 2021].

[7] National Geographic. Famine; 2020, https://www.nationalgeographic.org/encyclopedia/famine/. [Accessed 30 September 2021].

[8] Tulane University, School of Social Work. A closer look at the long-term health consequences of natural disasters 15 Mar 2018; 2018, https://socialwork.tulane.edu/blog/health-consequences-natural-disasters#:~:text=Outbreaks%20of%20malaria%2C%20measles%2C%20and,term%20effects%20of%20natural%20disasters. [Accessed 30 September 2021].

[9] World Health Organization. Ebola virus disease; 2021, https://www.who.int/health-topics/ebola#tab=tab_1. [Accessed 30 September 2021].

[10] Liang ST, Liang LT, Rosen JM. COVID-19: a comparison to the 1918 influenza and how we can defeat it. Postgrad Med J 2021;97(1147). https://pmj.bmj.com/content/97/1147/273. [Accessed 30 September 2021].

[11] Our World in Data. COVID-19 data explorer; 2021, https://ourworldindata.org/explorers/coronavirus-data-explorer?Metric=Confirmed+deaths&Interval=Cumulative&Relative+to+Population=true&Align+outbreaks=false&country=~OWID_WRL. [Accessed 30 September 2021].

[12] Center for Disease Control and Prevention (CDC). An introduction to applied epidemiology and biostatistics. In: Principles of epidemiology in public health practice. 3rd ed; 2012 [Accessed 30 September 2021] https://www.cdc.gov/csels/dsepd/ss1978/lesson1/section11.html.

[13] World Health Organization. Emergency: coronavirus disease (COVID-19) pandemic; 2021, https://www.who.int/. [Accessed 30 September 2021].

[14] Association for Professionals in Infection Control and Epidemiology. Outbreaks, epidemics and pandemics—what you need to know, https://apic.org/monthly_alerts/outbreaks-epidemics-and-pandemics-what-you-need-to-know/. [Accessed 30 September 2021].

[15] Center for Disease Control and Prevention (CDC). Epidemiology and prevention of vaccine-preventable diseases: tetanus; 2019, https://www.cdc.gov/vaccines/pubs/pinkbook/tetanus.html. [Accessed 30 September 2021].

[16] Center for Disease Control and Prevention (CDC). Chickenpox (varicella): outbreak identification, investigation, and control; 2021, https://www.cdc.gov/chickenpox/outbreaks.html. [Accessed 30 September 2021].

[17] Center for Disease Control and Prevention (CDC). Pandemic (H1N1 virus); 2019, https://www.cdc.gov/flu/pandemic-resources/1918-pandemic-h1n1.html. [Accessed 30 September 2021].

[18] Center for Disease Control and Prevention (CDC). Severe acute respiratory syndrome (SARS); 2017, https://www.cdc.gov/sars/about/fs-sars.html. [Accessed 30 September 2021].

[19] History.com Editors. World Trade Center is bombed. A&E Television Networks; 23 August 2021. https://www.history.com/this-day-in-history/world-trade-center-bombed. [Accessed 30 September 2021].

[20] Federal Bureau of Investigation. Oklahoma city bombing, https://www.fbi.gov/history/famous-cases/oklahoma-city-bombing. [Accessed 30 September 2021].

[21] History.com Editors. World Trade Center. A&E Television Networks; 10 September 2019. https://www.history.com/topics/landmarks/world-trade-center. [Accessed 30 September 2021].

[22] Cox JW, Rich S. As school shootings surge, a sixth-grader tucks his dad's gun in his backpack: the pandemic waned, classrooms reopened and gun violence soared at the nation's primary and secondary schools. The Washington Post 2021;(24 June). https://www.washingtonpost.com/education/2021/06/24/school-shootings-2021-increase/. [Accessed 30 September 2021].

[23] Cisco. What is a cyberattack? 2021, https://www.cisco.com/c/en/us/products/security/common-cyberattacks.html#~types-of-cyber-attacks. [Accessed 30 September 2021].

[24] Secureworks. Cyber threat basics, types of threats, intelligence & best practices; 2021, https://www.secureworks.com/blog/cyber-threat-basics. [Accessed 30 September 2021].

[25] Federal Emergency Management Agency; 2021, Theory, principles and fundamentals of hazards, disasters, and U.S. Emergency Management. https://training.fema.gov › aemrc › courses › hazdisusems [Accessed 30 September 2021].

[26] Federal Emergency Management Agency. The historical context of emergency management [chapter 1] https://training.fema.gov› hiedu › docs› chapter1 [Accessed 30 September 2021].

[27] Edwards-Winslow F. Changing the emergency management paradigm: a case study of San Jose, California, FEMA Training. https://training.fema.gov › hiedu › practitioner [Accessed 30 September 2021].

[28] United States Nuclear Regulatory Commission. Backgrounder on the Three Mile Island accident; 2018, https://www.nrc.gov/reading-rm/doc-collections/fact-sheets/3mile-isle.html. [Accessed 30 September 2021].

[29] Federal Emergency Management Agency. Strategic plan 2008-2013, FEMA P-422; 2008, https://www.fema.gov/pdf/about/fy08_fema_sp_bookmarked.pdf. [Accessed 30 September 2021].

[30] Drabek TE, Hoetmer GJ. Emergency management: principles and practice for local government. Washington, DC: ICMA; 1991.

[31] Selves MD. Local emergency management: a tale of two models. ASPEP J 1997;63–7.

[32] Blaikie P, Cannon T, Davis I, Wisner B. At risk: natural hazards, people's vulnerability, and disasters. London: Routledge; 1994.

[33] Hammer CC, Brainard J, Innes A, Hunter PR. (Re-) conceptualising vulnerability as a part of risk in global health emergency response: updating the pressure and release model for global health emergencies. Emerg Themes Epidemiol 2019;16(2). https://doi.org/10.1186/s12982-019-0084-3 [Accessed 30 September 2021].

[34] Etkins D. Disaster models. In: Butterworth-Heinemann, disaster theory: an interdisciplinary approach to concepts and causes. Elsevier; 2016. p. 193–228. https://doi.org/10.1016/B978-0-12-800227-8.00006-5 [Accessed 30 September 2021].

[35] Canadian Standards Association. Risk management: guideline for decision makers; 2002, http://lib.riskreductionafrica.org/bitstream/handle/123456789/743/risk%20management.%20guidelines%20for%20decision-makers.%20a%20national%20standard%20of%20canada.pdf?sequence=1. [Accessed 30 September 2021].

[36] CARE. Household livelihood security assessments: a toolkit for practitioners; 2002, https://www.careemergencytoolkit.org/. [Accessed 30 September 2021].

[37] Morris ZA, Hayward RA, Otero Y. The political determinants of disaster risk: assessing the unfolding aftermath of Hurricane Maria for people with disabilities in Puerto Rico. Environ Justice 2018;11(2):89–95.

[38] Sexton K, Hattis D. Assessing cumulative health risks from exposure to environmental mixtures—three fundamental questions. Environ Health Perspect 2007;115(5):825–32.

[39] Erickson W, Lee C, von Schrader S. Disability statistics from the 2018 American Community Survey (ACS); 2021, www.disabilitystatistics.org. [Accessed 30 September 2021].

[40] Tarrant M. The organization: risk, resilience and governance. Aust J Emerg Manag 2010;25(2):13–7.

[41] Jardine C, Hrudey S, Shortreed J, Craig L, Krewski D, Furgal CM, McColl S. Risk management frameworks for human health and environmental risks. J Toxicol Environ Health B 2003;6(6):569–720.

[42] Tarrant M. Regional workshop on total disaster risk management. Emergency Management Australia (EMA); 2002. http://www.adrc.asia/publications/TDRM/17.pdf. [Accessed 30 September 2021].

[43] Federal Drug Administration. Current and resolved drug shortages and discontinuations reported to FDA; 2021, https://www.accessdata.fda.gov/scripts/drugshortages/. [Accessed 30 September 2021].

[44] McLoughlin D. A framework for integrated emergency management. Public Adm Rev 1985;45:165–72 [Special Issue].

[45] National Governor's Association Center for Policy Research. Comprehensive emergency management: a governor's guide; 1979.

[46] Tuscaloosa EMA. Tuscaloosa county emergency management cycle; 2003, https://www.tuscaloosacountyema. org/about/evolution-of-emergency-management. [Accessed 30 September 2021].

[47] Manitoba Health. Disaster management model for the health sector, guideline for program development., 2000, https://vdocuments.net/disaster-management-model-for-the-health-manitoba-health-disaster-management.html. [Accessed 30 September 2021].

[48] The National Archives. Integrated emergency management and key organizations; 2005, https://webarchive.nationalarchives.gov.uk/20050112000000/http://www.ukresilience.info/contingencies/dwd/c2amanagement.htm. [Accessed 30 September 2021].

[49] Nojavan M, Salehi E, Omidvar B. Conceptual change of disaster management models: a thematic analysis. Jàmbá 2018;10(1):1–11. https://doi.org/10.4102/jamba.v10i1.451 [Accessed 30 September 2021].

[50] HM Government. UK emergency planning: our context and approach, https://www.oasis-open.org/committees/download.php/55686/Overview_of_UK_Emergency_Management_vers1.0.pdf. [Accessed 30 September 2021].

[51] Cuny F. Principles of management: introduction to disaster management. Prehosp Disaster Med 1998;13(1):80–5.

[52] Cuny FC. Training syllabus for the UNHCR emergency managers training workshop Madison Wisconsin. University of Wisconsin Disaster Management Center; 1987. https://oaktrust.library.tamu.edu/bitstream/handle/1969.1/159931/cuny_intertect_000002-40.pdf?sequence=1&isAllowed=y. [Accessed 30 September 2021].

[53] Abuja ATN. Remembering Nigeria's Biafra war that many prefer to forget; 2020, https://www.bbc.com/news/world-africa-51094093. [Accessed 30 September 2021].

[54] Federal Emergency Management Agency. Principles of emergency management supplement; 2007, https://www.fema.gov/media-librarydata/20130726-1822-25045-7625/principles_of_emergency_management.pdf. [Accessed 30 September 2021].

[55] Tulchinsky TH, Varavikova EA. A history of public health. In: The new public health; 2014. p. 1–42. https://www.ncbi.nlm.nih.gov/pmc/articles/PMC7170188/. [Accessed 30 September 2021].

[56] Rhodes P, Bryant JH. Public health. In: Encyclopedia Britannica; 2021. https://www.britannica.com/topic/public-health. [Accessed 12 July 2021].

[57] Rose DA, Murthy S, Brooks J, Bryant J. The evolution of public health emergency management as a field of practice. Am J Public Health 2017;107(Suppl. 2):S126–33.

[58] World Health Organization. Preamble to the constitution of the World Health Organization as adopted by the International Health Conference. New York, 19-22 June, 1946; signed on 22 July 1946 by the representatives of 61 States (Official Records of the World Health Organization, no. 2, p. 100) and entered into force on 7 April; 1948.

[59] Center for Disease Control and Prevention (CDC). Geography and geospatial science working group (GeoSWG); 2019, https://www.cdc.gov/gis/index.htm. [Accessed 30 September 2021].

[60] Center for Disease Control and Prevention (CDC). Fact sheet: medical countermeasures (MCM) and points of dispensing (POD) basics; 2020, https://www.cdc.gov/cpr/readiness/healthcare/closedpodtoolkit/factsheet-mcm.htm. [Accessed 30 September 2021].

[61] Buckler P, Mars G, Smale S. New approaches to assessing vulnerability and resilience. Aust J Emerg Manag 2000;(Winter):8–14.

[62] Smith F, Jolley E, Schmidt E. Disabilities and disasters: importance of an inclusive approach to vulnerability and social capital. Sightsavers, United Nations Development Group; 2012. p. 1–17. https://www.preventionweb.net/publication/disability-and-disasters-importance-inclusive-approach-vulnerability-and-social-capital. [Accessed 30 September 2021].

[63] Fjord L, Manderson L. Anthropological perspectives on disasters and disability: An introduction. Hum Organ 2009;68(1):64–73.

[64] Krahn GL, Walker DK, Correa-De-Araujo R. Persons with disabilities as an unrecognized health care disparity population. Am J Public Health 2015;105(Suppl. 2):S198–206.

[65] Kett M, Twigg J. Disability and disasters: towards an inclusive approach. In: IFRC, world disasters report. Focus on Discrimination; 2007. p. 86–111.

[66] Priestley M, Hemingway L. Natural hazards, human vulnerability and disabling societies: a disaster for disabled people? Rev Disabil Stud 2014;2(3):1–14.

[67] Martin SA. A framework to understand the relationship between social factors that reduce resilience in cities: application to the City of Boston. Int J Disaster Risk Reduct 2015;12:53–80.

[68] National Health Preparedness Index; 2021, https://nhspi.org/about/. [Accessed 30 September 2021].

Populations with disabilities and others with access and functional needs

Jill Morrow-Gorton

University of Massachusetts Chan Medical School, Worcester, MA, United States

Learning objectives

(1) Describe the populations of people with disabilities that have access and functional needs.

(2) Discuss factors leading to health inequities, including social determinants of health and race and analyze how these impact populations of people with disabilities.

(3) Explain the impact of the co-occurrence of multiple factors on people with disabilities.

Integrating Mental Health and Disability Into Public Health Disaster Preparedness and Response
https://doi.org/10.1016/B978-0-12-814009-3.00005-2

Introduction

Effective emergency planning requires understanding the populations of people that will be affected by a potential disaster and the disaster's impact on the physical and social infrastructure. Historically, some populations of individuals have not been incorporated into disaster and emergency planning, thus creating a system that is unable to respond effectively to their specific needs. In addition, identifying where these populations may be found is critical to the ability to plan for adequate resources to respond in the event of a disaster. Demographics differ between municipalities, states, regions, nationally and internationally, making it imperative for emergency planners and public health officials to understand the populations with which they are planning. Understanding the characteristics and profiles of the community's populations, including their past disaster experiences, where they live in the community, how they get around, and what support resources they have will impact how to plan with and for them. The principle of "all disability" should be considered like that of "all hazard" to reflect the needs of all people with functional needs impacting disaster preparedness when planning with any community. This chapter focuses on defining and describing the populations of people with disabilities and chronic physical and mental health conditions in order understand their needs before, during and after a disaster.

Language

Spoken and written language as a means of communication is characterized by the use of words in a structured, conventional manner [1]. This communication may be specific to a particular social group or more broadly accepted by a larger population of people. It is used to describe people and places, express individuality and imagination and is integral to the identity of a population. All language evolves over time and language used to refer to people with disabilities is no different. Even the expression "disability" is no longer universally agreed upon. In addition, how individuals or groups of individuals with similar characteristics refer to themselves, may not reflect how they prefer to be referred to by those outside of their group. While individuals and particular groups may have their preference for terms, the principles of treating all people with respect and dignity and making them feel comfortable and accepted are paramount.

Language and how one refers to oneself is central to one's being and identity. Disability language evolved across the 20th century in attempts to remove the stigma from language. Preferences about language are not universal either between or within disability groups with some choosing "person-first" language and others preferring "identity-first" language. People choose their own identity and their choice of language reflects that regardless of which language they prefer. Preferences regarding language often, but not always differ by disability group.

"Person first language" became one means of referring to people with disabilities in order to think about them as a person rather than a disability or diagnosis and to counteract the use of more pejorative terms [2,3]. In person-first language, the sentence structure reflects the person first and then their disability. Thus, one would say a "woman with a disability" instead of a "disabled woman." While the disability advocacy movement in the United States began in the 1970s, grounded in the civil rights movement of the 1960s, and included the passing of the Rehabilitation Act in 1973, advocates did not begin using the term "person-first" language until

1980s or early 1990s [4,5]. Advocacy organizations for people with disabilities, including intellectual and developmental disabilities, such as the ARC and Speaking for Ourselves espouse person-first language noting that people with disabilities are people with individual interests beyond their disability [6,7]. These groups promote self-advocacy and self-determination on the part of individuals with intellectual and developmental disabilities as well as advocacy for parents, families and communities.

Alternative preferences exist and some individuals and groups adopt the concept of "identity-first" language which emerged as a movement among some disability communities as a way to counteract stigma. These individuals and communities choose to identify themselves by their disability as one would identify by any other characteristic. They call themselves disabled and just as one would be called American [8]. "Identity-first" language reflects that disability is part of who the person is and a cultural identifier. "Identify-first" language is generally preferred by self-advocates in the autistic, deaf, and blind communities [9–13].

In 1993 the National Federation of the Blind (NFB), the oldest and largest organization led by blind people in the United States, reflecting the collective views of its members in noting that blindness is just one characteristic for people with limited or no vision and like other characteristics is more efficiently put in front using "blind person" rather than "person with blindness" [9]. The Deaf community has also rejected person-first language preferring instead "deaf person" or "hard of hearing person" [10]. In autism advocacy, there is somewhat of a split between "person-first" and "identity-first" language. Some self-advocates with autism including the Autistic Self-Advocacy Network (ASAN) prefer "identity-first" language and refer to themselves as autistic. Others, such as the autism advocacy group Autism Speaks use "person-first" language, a person with autism [11–13]. This dichotomy in preferences makes determining which construct to use somewhat difficult. However, when working with a particular person or group meeting their preference is important.

Access and functional needs

A broad construct for considering populations needing special consideration in disaster planning has been posed for use in the public health and emergency management sectors using the term "access and functional needs" or AFN [14]. In the past, including post Hurricane Katrina, the term "special needs" has been used to consider these populations. "Special needs" is an amorphous term with numerous definitions referring to a broad range of people with various kinds of needs. In 2004 the Centers for Disease Control and Prevention, CDC, included in the definition minority groups, non-English speakers, children, the elderly and people with serious mental illness in addition to people with disabilities [15]. Additionally, the Federal Emergency Management Agency (FEMA) and others have added other groups to that list including single working parents, people without vehicles, people with special dietary needs, pregnant women, prisoners, and people who are homeless [14]. The impact of Hurricane Katrina on individuals with disabilities and many others considered to be people with special needs led to recommendations that emergency and public health preparedness planners take these population's needs into account in emergency planning. With this, the definition of special needs came under scrutiny in order to better identify the populations and their emergency planning needs. Disability experts and advocates, Kailes and Enders [16], began the conversation by rejecting the notion of special needs emergency planning, as too broad and too difficult to operationalize for emergency planning and response. They argued that considering the functional needs of individuals with disabilities and others traditionally encompassed under the "special needs" label makes for more efficient and effective planning. In this manner, similar needs of different populations could be tackled together, simplifying the process and the response. As the needs of these populations vary both between and within the groups, a new term, "access and functional needs," was coined to reflect those similar functional needs rather than distinct populations. Thus, emergency management and public health replaced "special needs" with AFN as a way to better understand the emergency needs for these populations.

Characterization of the terms "access needs" and "functional needs" furthers the exploration of these concepts. Irrespective of specific diagnosis, status, or label, access-based needs impact the ability to acquire certain resources, such as social services, accommodations, information, transportation, medications to maintain health, and others. "Functional needs" refer to task-based limitations resulting in an individual requiring additional support before, during, and after a disaster or emergency [17]. An example of a functional need would be someone who is Deaf and can't hear a tornado warning siren. This person has a functional need related to communication as they can't hear. This person also has an access need to gain access to a tornado warning. Therefore, the warning must be communicated via a method that the person can access, such as a text.

The term "access needs" refer to the challenges that various populations experience in order to participate in the community every day which become magnified in the event of an emergency or disaster. Obstacles include access to medical care and medications, transportation, food, and other life essentials. In addition to populations that experience these access challenges on a regular basis, other populations, described as "at risk" individuals may not face these difficulties every day, but confront them during an emergency or disaster. The Pandemic and All-Hazards Preparedness Reauthorization Act of 2013 defines "at-risk" individuals as individuals, such as children, older adults, pregnant women, and individuals who may need additional assistance during an emergency or disaster response [17]. The US Department of Health and Human Services (HHS) further considers where at-risk individuals live, such as in the community, in institutions, or whether they are homeless, to be important at-risk factors. Other at-risk factors include the cultural and linguistic diversity of populations, including people from diverse cultures, with limited or no English proficiency. Finally, the Department adds individuals with chronic medical conditions including substance use disorders as well as disabled individuals.

The CMIST framework advocated for by HHS provides a basis for concisely considering the needs of populations with access and functional needs, including those at-risk individuals [18]. The acronym CMIST references the five broad categories of functional need common to people with AFN and help emergency management and healthcare providers address those needs in planning and in an emergency. The five CMIST categories include Communication, Maintaining health, Independence, Support, and Transportation. Expanding the individual concepts in the CMIST Framework helps explain the important aspects of each.

Communication is critical to effective emergency planning, response and recovery and an area where people with access and functional needs often require accommodations or "modifications" as specified under the Americans with Disabilities Act. Many individuals have limitations that interfere with the receipt of and response to information when presented as it is to the general public. These include a broad range of people, some with vision, hearing or cognition disabilities and others with limited English proficiency [18]. For people with cognitive disorders, such as intellectual disabilities or dementia and those with limited or no English proficiency, accessing information in English in any form may be problematic. The Deaf and Hard of Hearing population that uses sign language to communicate may also be challenged by certain forms of communication. The accessibility needs of various populations can affect all types of communication, including verbal, written, and visual. Some people require more than one communication accommodation, for example, needing large print text plus simplified text. Consideration for making communication accessible to all is critical to emergency planning and response.

Verbal communication generally involves either an in-person dialog between two or more people or a conversation using technology, such as a phone or computer. During emergencies, messages communicated to the public in affected areas come in a variety of forms including voice or text messages to the phone, email, loudspeaker announcements or sirens to alert people to the need to evacuate or take other action or the presence of first responder personnel. Many of these methods of communication may be inaccessible to Deaf and Hard of Hearing individuals who have a range of hearing levels from some hearing to hearing only at certain frequencies to no hearing. As well, they may communicate using a number of different methods including verbal speech, lip reading and sign language. However, even sign language is not universal as British Sign Language and American Sign Language are not the same and people that use one may not be able to understand those that use the other. People who have hearing disabilities may not be able to access some of the standard verbal voice messages or announcements and need adaptive means, such as a video relay system or a TTY that uses text for back and forth communication [19]. In addition, they may use modifications, such as flashing lights for sounds like the doorbell in order to alert them that someone is at the door.

Accessing written language can be an issue for many individuals. People who are Blind or visually impaired or those with low vision have varying levels of ability to read written materials. Some may be able to read large print, while others may not. Computer screens or smart phones can be fitted with screen reader software to access written materials on-line in lieu of paper materials. Screen readers read aloud the words on the computer or phone screen to the person. Individuals with cognitive disorders and dementia may not be able to access written documents as they may not be able to read. For them, visual pictures can be used to transmit the message. Documents translated into languages other than English can help meet the needs of those with limited English proficiency. Consequently, in emergency planning it is important to take all of these potential needs into account in order to be able to reach all of the people in the community. In particular, it is important to understand the specific needs of the people within the local community so that the focus of resources is on creating communication that will reach them.

Maintaining health, the second element of the CMIST framework, refers to services and supports that help keep people healthy and manage their chronic medical conditions. Some of these services address physical health conditions, such as diabetes or asthma and include having available the appropriate medications to treat the condition. Many individuals with multiple chronic conditions may take a number of medications or require treatments to manage their health. For example, some with chronic lung disease are dependent on oxygen to breathe effectively. People who do not walk independently may use a wheelchair to get around. For those using a power wheelchair that runs on batteries requiring regular charging, access to electricity to keep the batteries charged and to an environment where they are able move without stairs or other barriers is crucial. Others require assistance to complete activities of daily living including eating, dressing, grooming, transferring or moving from one space to another, and toileting. These individuals often have caregivers that may live with them or come to their home to help with these tasks. They may require equipment for transfers and special nutrition to support tube feeding if unable to eat regular food. All of these needs must be considered when planning for evacuation and sheltering of these individuals in order to keep them healthy during an emergency.

Independence represents another critical aspect of supporting people with access and functional needs. The concept requires consideration of how to support, promote and maintain an individual's ability to function independently in a different environment or their own home despite a disaster. Many individuals with disabilities use medical equipment and assistive devices in order to function optimally on a day to day basis. The nature of these devices may differ. Some such as power wheelchairs, walkers, and scooters support physical needs, while others help meet communication needs like alternative and augmentative communication (AAC) devices. AAC devices can be low tech communication books or high tech, computerized aids and equipment that facilitate communication for those who cannot produce comprehensible speech or written language or in some cases which face challenges in understanding directions to complete a task. Electronic equipment and devices must be charged and maintained. Consumable medical supplies including diapers or briefs, formula and tube feeds, bandages, ostomy supplies, and insulin syringes meet other physical needs of many. These products are distinguished by their non-durability, inability to be reused, and disposability. They must be replenished regularly because they cannot be reused. Items, such as syringes and needles also require safe disposal because of the potential for accidental injury to others and exposure to potential infections. Service animals represent another type of support used by people with disabilities to help preserve independent functioning. In order to maintain independence, people that use these devices, products, and supports must have constant access to them and the ability to maintain them in good working order during and after a disaster.

In addition to devices and supplies, some people require support in the form of supervision to remain safe. People with cognitive disabilities including intellectual disabilities and dementia may not independently have the judgment or understanding to make decisions or assess the level of risk related to particular activities. Consequently, they may need to have someone support them in making these safety-related decisions. It should be noted, however, that many people with these conditions may be able to make decisions. It should not be automatically assumed that someone with a particular disability can or cannot make decisions independently. Other people including those with serious mental illness may exhibit a range of abilities to function. Some people cope well in the absence of an emergency and have the capability of making decisions independently, while others may require support or assistance in day to day life. During emergencies even individuals with independent functioning may need assistance and others may require more support or assistance than before the emergency. One of the World Health Organization (WHO) keys to promoting recovery is supporting people, especially those with mental health disorders, so that they feel safe, connected, and calm, and can figure out how to remain functional [20]. Good mental health and mental health support promotes wellbeing and functioning that are crucial to individual and community resilience. Other populations, such as pregnant women, nursing mothers, and infants, and children may also have a temporary need for additional support during and after a disaster in order to maintain their wellbeing.

FEMA's Functional Needs Support Services (FNSS) are services that enable individuals with access and functional needs to maintain their independence in a general population emergency shelter, known as a "mass care" shelter [21]. Examples of shelter support services include durable medical equipment (DME), consumable medical supplies (CMS), and personal assistance or personal care services (PAS or PCA). Other needs may include communication assistance and services. Availability of food and beverages appropriate for individuals with dietary restrictions or those that take in nutrition through an enteral tube or a central venous line is also critical.

Emergency planners and public health officials need to know and understand the community's demographic profile in order to determine what supports, services and equipment will be needed in a disaster or emergency. Meeting with community partners, stakeholders, providers, constituents, and service recipients, including individuals with access and functional needs, will enhance emergency planners' and public health officials' abilities to develop plans that successfully integrate individuals with access and functional needs into general population shelters. In addition, these collaboration efforts will help educate community members with access and functional needs about the importance of personal preparedness plans.

The final element of the CMIST framework, transportation, affects a broad range of people based on differing needs. Many people with disabilities require and use accessible public or private transportation to reach goods, services and activities [22,23]. Accessibility refers not only to the physically accessible transportation, but also to the cost and time factors involved. People with disabilities need to get to all of the places that people without disabilities do including work, school, shopping, entertainment, medical care, and others. Accessible transportation must meet the communication needs of people who are Blind or have visual impairments as well as those that are Deaf and Hard of Hearing. It needs to accommodate various methods of ambulation including wheelchair use, walkers, and crutches. It also has to meet the needs of people with access and functional needs who may not consider themselves disabled. For example, an older adult who walks with a cane because of arthritis and a person with cerebral palsy who uses crutches both need assistance getting up the stairs to enter a bus. In both examples, low floor or low entry buses may be useful. These buses remove the need for climbing stairs by bringing the bus floor down to the street level, but they also may assist children, people with short legs, people carrying bundles and other people without disabilities. Thus, the modifications and accommodations available in transportation can benefit the community in general in addition to making transportation accessible to those with disabilities and motor needs.

During disasters, however, transportation availability can become tenuous and public transportation may be disrupted impacting the general public, including people with disabilities. Additional groups which may be affected include those with limited incomes who may have a vehicle, but may not be able to afford the expenses of gas and tolls to evacuate out of their home area. Planning for disasters requires consideration of the many factors involved with providing accessible transportation for all of those in the community who may not be able to evacuate independently and coordination to make sure they can access it.

Focusing on access and functional needs rather than individual disability type, diagnosis, status or label allows emergency planners to incorporate a broad set of common needs into the planning process. Organizing these needs into the CMIST framework helps planners at all levels use a common language and common elements when thinking about these needs. It also permits them to develop a more efficient, streamlined emergency plan and structure. Thus, all federal, territorial, tribal, state, and local emergency and disaster plans can contain a description of the strategies to address access and functional needs in a concise and comprehensible way, with clear instructions for implementation. This efficiency not only assures that the information is there and considered, but also simplifies training and response as it avoids the need for multiple documents and multiple approaches. As well, it integrates these resources into the plan so that they can be harnessed for all "at-risk" individuals regardless of whether those needs are temporary or permanent. However, while CMIST provides the framework, it must be recognized that not all people with disabilities have the same access and functional or other needs and that with the exact same diagnosis might have very different access and functional needs. The CMIST framework helps make planning for community needs more predictable and comprehensive but it can't account for unexpected occurrences or needs during an actual event.

Demographics of people with access and functional needs

Populations of people with access and functional needs include a wide array of people, comprising those with disabilities, older people, children, people with limited English proficiency, and in the case of transportation, those dependent on public transportation which may not function during a disaster. In the United States and other countries, functional needs are measured using six questions developed by the Washington Group using the World Health Organization's (WHO) International Classification of Functioning, Disability, and Health (ICF) which defines disability for the purposes of public health surveillance [24,25]. These differ somewhat from the CMIST categories and capture the percentage of adults with: (1) serious difficulty walking or climbing stairs; (2) serious difficulty concentrating, remembering or making decisions; (3) difficulty completing errands independently; (4) Deafness or serious difficulty hearing; (5) blindness or serious difficulty seeing; and (6) difficulty bathing or dressing. People who have difficulties with independent mobility or cognitive functions encompass 13.6% and 10.8% of the adult population, respectively. The other functional disabilities are present in the following percentages of the adult population: independent living, 6.8%; hearing, 5.9%; vision 4.6%; and self-care 3.7%. There is also overlap between categories, and those numbers reflect the overlaps. The most common overlap is found in people with multiple functional disabilities include mobility, cognition and who face challenges with independent living. Stevens and colleagues note that about 60% of adults have only one disability, based on answers to those six questions [26]. Of the remaining 40% of adults with disabilities, the percentages of those that have multiple functional disabilities are 21% with two, 11.9% with three functional needs, and 9.7% with four or more functional needs.

Adults with functional disabilities also have more health and other disparities as compared to the general public. In general, they have a lower socioeconomic status (SES) with the number of low SES indicators increasing as the number of functional disabilities increases [26]. They are less likely to have a high school education and more likely to be out of work and looking for work. People with disabilities are also more likely to have an income-to-poverty ratio of less than one, another indicator of low income.

Social determinants related to poverty, for example, unstable housing and food insecurity have been shown to have a significant impact on health resulting in poorer health outcomes. A study of breast cancer outcomes by geographical areas related to poverty in New Jersey indicated decreased breast cancer survival rates in poorer neighborhoods [27]. Studies of the impact of poverty on diabetes in California indicate that people with diabetes living in poverty are more likely to have a limb amputated than those not in poverty [28]. Other health disparities more frequently present in people with functional disabilities include an increased rate of obesity, smoking, heart disease and diabetes [29]. As well, they have less access to and use of healthcare services; one in three people with disabilities report lacking a primary care provider and the presence of a healthcare need. Only 25% of this group has also had a routine annual check-up in the last year compared to 62% of the general population [29,30]. Thus, these populations experience disparities in both socioeconomic status and health related indicators.

In the United States differences exist in the populations of people with disabilities related to where they live, race and ethnicity, and other characteristics. The prevalence of disability in the US population varies based on the definitions used for disability. A study by the Pew Research Center using census data showed that in 2015 nearly 40 million Americans with a disability lived in the United States, representing 12.6% of the civilian non-institutionalized population [31]. The CDC data shows a prevalence of 26% based on data from their Disability and Health Data System (DHDS) [29]. Geographically the prevalence of people with disabilities varies not only by state, but also by localities within states. While in general disabilities are most prevalent in the south, pockets of states with high disability rates occur throughout the country. Arkansas, Kentucky and Alabama represent states with some of the highest level of disability, around 17%. West Virginia has the highest rate in the country at 19%, contrasting with Utah with the lowest rate of 9.9%. In Kentucky, the disability prevalence of 28.7% in Pike County greatly exceeds the 17% rate for the state as a whole. In other states with lower statewide disability prevalence, particular towns and cities have significantly higher rates of disability. For example, disability prevalence in Pueblo, Colorado is 22%, but just 10.3% statewide. On the other hand, Indiana has a slightly higher statewide prevalence of about 13.8% with a low prevalence in one of its cities, Fisher, of 3.5%. This variation makes it critical for emergency planners to know and understand the demographics of the area in which they work in order to best meet the needs of the local population.

History has revealed that disasters result in disparate outcomes in different populations. The evidence shows that populations of people with access and functional needs experience more negative outcomes with longer lasting impact compared to the general population. Actively engaging these populations in disaster planning and understanding them better will help mitigate those negative impacts by improving response for the disability population and empowering them to become active participants, thus increasing their resilience. It is important to understand who these populations are and where they can be found in order to engage them in a meaningful way and determine how best to plan disaster response and recovery with them. Before looking at the details of each group, there are principles to review that impact all groups.

Prior to the 1999 US Supreme Court's decision in *Olmstead v. L.C.,* many people with disabilities, including older people, were segregated into institutional settings such as nursing facilities and developmental centers [32]. Even though some states have made significant progress in closing these institutions and providing community living opportunities for people through home and community based waivers, there remain variations between states regarding the proportion of funding for community based services and those provided for institutional living [33]. While this likely will continue to change over time, it illustrates how individual areas differ in ways that may have dramatic impacts on emergency planning and response. What follows is an in-depth look at each of these populations to give some perspective about how to think about planning for and with them.

People with physical disabilities

The population of people with physical disabilities represents one of the most diverse groups in terms of age, gender, and other characteristics and different conditions leading to functional disability. Physical disabilities that affect physical functioning most commonly affect mobility. In the United States it is estimated that 1 in 5 people have a disability and that 1 in 8 have a mobility issue [25]. These numbers vary from state to state, with Minnesota having the lowest number of people with disabilities (16.4%) and Alabama having the highest number (31.5%) in 2013. People with physical disabilities face barriers to movement in everyday life. Wheelchairs can't climb stairs. People who walk with supports, such as canes or crutches, may be able to climb a few stairs but during an emergency are unlikely to be able to maneuver the multiple flights of stairs in a tall apartment or office building. Causes of physical disabilities include acquired or congenital conditions that affect physical functioning. Acquired conditions vary greatly in terms of the populations that they affect and their impact on function. Usually these conditions alter neurologic functioning or that of muscles and bones and have a variety of etiologies. Each may have a different kind of impact, but the functional disruption is likely similar. Congenital conditions, while different, impact functioning in a similar way.

Examining the etiologies of conditions leading to mobility limitations will give a perspective about the conditions themselves and build a better understanding of the needs associated with them. In older adults, osteoarthritis is one of the top two conditions leading to decline in physical function and increase in physical disability [34]. The other condition at the top of the list, cardiovascular disease, has long been known to contribute to decreased mobility related to symptoms that limit movement such as chest pain and difficulty breathing [35,36]. Stroke, also related to cardiovascular disease, affect physical functioning to varying degrees depending on the extent of the stroke damage and recovery. Estimates are that more than half of stroke survivors over the age 65 have reduced mobility post stroke. In the general population, functional mobility impairment associated with paralysis has multiple causes and two-thirds of the people affected are between the ages of 18 and 64 years. The 2013 US Paralysis Prevalence & Health Disparities Survey analyzed the prevalence and causes of functional limitations from paralysis affecting an estimated 5.4 million people in the United States [37]. While stroke topped the list at 33.7%, it was followed closely by spinal cord injury in 27.3% of the population with paralysis, 18.6% of people with multiple sclerosis, and 8.3% with cerebral palsy. The survey identified many other causes, however, at much lower and more inconsistent frequencies. Although the underlying etiologies for these conditions differ, ranging from developmental conditions like cerebral palsy to accident-related spinal cord injury, the presence of functional mobility limitations links them to a common planning strategy to support those limitations in the event of an evacuation.

In general, the proportion of people with disabilities rises with age. According to the 2017 annual report of Disability Statistics and Demographics, ambulatory disabilities affect 0.6% of children and youth 5 to 17 years of age, 5.1% of adults 18 to 64 years old, and 22.5% of adults 65 years and over [38]. Of individuals 18 to 64 years, 3.8% are unable to live independently, while this applies to 14.6% of those 65 years and older. The proportion of people with ambulatory disabilities has remained relatively stable from 2008 to 2016 with only a small rise of 0.2% over that time. Employment rates differ significantly for those with ambulatory disabilities compared to those without. About 35.9% of people with disabilities between the ages of 18 and 64 years were employed, in contrast to 76.6% of those in the same age group without disabilities. Earning disparities between these groups exist as well. From 2008 people with disabilities earned on average of about $10,000 less a year than those without, and that gap has widened since 2013 [31,39]. In 2016 people with disabilities earned about two-thirds of that of those without disabilities. Their poverty rate between 2009 and 2016 was 22% compared to 14.3% in the non-disabled population [40]. The literature shows that the poverty percentage gap or the difference between the percentages for people with disabilities versus those without is on average 7.7 points. Thus, poverty is a significant issue for people with disabilities.

Housing also poses a significant issue for people with disabilities especially those who are not elderly While there are programs for housing seniors, those may not be available to younger people with disabilities. As well, in absolute numbers, the population of people 35 to 64 years with disabilities exceeds that of any other age group, making the housing problems for younger people with disabilities even more difficult [40]. Poverty also plays a big role in where people with disabilities live. Compared to those without disabilities they tend to live in less desirable neighborhoods with poorer access to public services, lower median incomes, higher crime rates, and other neighborhood problems like noise

[41]. Lack of accessible housing also impacts where people with disabilities live, often obliging them to live in inadequate, overcrowded housing [42]. In addition to lower incomes, people with disabilities are more likely to pay more than 50% of their income on rent, placing them even deeper into poverty [43]. The US Department of Housing and Urban Development (HUD) provides federal housing assistance for many people with disabilities. HUD data shows that income in HUD-assisted households with at least one person with a disability is the lowest, regardless of geography or size of the public housing agency. While there are housing programs specifically targeting people with disabilities and providing needed accessibility accommodations, many people with disabilities access HUD housing through programs that don't offer needed disability accommodations.

Health consequences of living in poverty are significant. Evidence shows that poor neighborhoods lack grocery stores and other resources to obtain healthy foods. They also lack the availability of fitness centers, local parks, bike paths, and other recreational features built into the environment to promote physical activity and healthy living [28,30,31]. These contribute to poorer health outcomes, with increased prevalence of obesity and associated comorbidities including hypertension and diabetes that accompany lack of spaces for physical activity. Rates of obesity in people with disabilities exceed those of the general population and in 2015 the rate of obesity in people with disabilities was 39.9%, compared to 25.4% in those without disabilities [44]. Obesity leads to associated health conditions such as high cholesterol which contributes to heart disease and stroke. People with disabilities have other health disparities, especially related to health behaviors. They are more likely to be smokers, 23.4% versus 14.9% in the general population, which contributes to chronic lung diseases, heart disease, stroke and lung cancer [45]. Chronic lung diseases such as Chronic Obstructive Pulmonary Disease (COPD) and emphysema put people at a higher risk for complications and death during influenza pandemics and disasters that increase particulate matter in the air including wildfires. Additionally, people with advanced lung disease may require treatment with oxygen, compounding their needs both at home and in the event that they evacuate. People with disabilities may have chronic health conditions that are a consequence of their disabling condition. For example, people with rheumatoid arthritis, an autoimmune disease causing inflammation that affects joints and mobility, also have an increased risk of heart disease related to the inflammation. Some developmental disabilities, such as cerebral palsy, are associated with a higher prevalence of sensory disorders including visual and hearing impairments that make mobility and communication more difficult.

People with mobility issues and health conditions often require equipment to support their functioning. Mobility devices such as canes, walkers, and rollators are commonly used by individuals with physical disabilities as are both manual and electric wheelchairs. Canes and walkers do not often require modifications to fit the user. Wheelchairs, on the other hand, may require significant customization to meet the user's individual needs. Form fitting seating, special steering devices such as joy sticks and other modifications are designed to meet the needs of the individual user and support their ability to move independently. People with disabilities affecting motor function may also use more than one mobility device. Some are able to walk short distances within their homes using a cane or walker, but require a wheelchair for mobility in the community for shopping and other activities.

Physical health conditions associated with disabilities may also require devices or special delivery systems for medication or other support. For people with COPD and oxygen dependency, they must wear their oxygen either all the time or at specified times during particular activities. People with sleep apnea may need to use a machine to help improve breathing during sleep. Continuous positive airway pressure (CPAP) and Bilevel Positive Airway Pressure (BiPAP) machines also need to travel with people when they are sleeping away from home. Others may use nebulizers to deliver aerosolized medication into the lungs to assist with respiratory conditions such as COPD and asthma. People with diabetes using insulin will need their supplies to inject their insulin regularly as well as access to refrigeration to maintain the insulin. Many other types of devices and supplies exist to support and treat health conditions and disabilities. In the event of a disaster and evacuation, it is important to remember to bring these with the evacuee they maintain health and will likely be difficult to replace in a shelter or temporary housing. Thus, supporting people with physical disabilities during a disaster must address more than just their physical disability and assure that they have the needed equipment to address mobility and associated chronic health conditions.

People with cognitive impairments: Overview

The Oxford Dictionary defines cognition as "the mental action or process of acquiring knowledge and understanding through thought, experience, and the senses" [46]. This includes executive functioning where a person uses information about a current situation, brings in knowledge and experience from the past, and is able to direct actions toward an abstract goal [47]. This executive control requires the ability to develop and choose from a number of solutions by predicting the possible outcomes of each and choosing the best one. Cognitive impairments impact the ability to think, remember, pay attention, make decisions and learn new things [48]. They also affect talking and interacting with others. These impairments have multiple causes and impact everyday functioning in various ways. Some people develop cognitive impairments through head injuries or medical conditions like dementia, while others are born with conditions that cause cognitive impairments like Trisomy 21 or Down syndrome. Congenital cognitive impairments, those present from birth or before, and those that occur in children before the age of about 18 years are called "intellectual disabilities" [49]. Cognitive impairments acquired in adulthood generally are referred to as "cognitive disorders."

Cognitive impairments may vary in severity and are categorized as mild, moderate, severe or profound, with profound primarily applying to intellectual disabilities [47,49]. Some people may have functional difficulties in only one or two areas of cognitive functioning such as in attention or learning disabilities while others have more affected areas. Cognitive disorders can occur with other conditions such as autistic disorder which involves social and communication difficulties as well as characteristic behavior and may be of any level of severity. Cognitive disorders also may be the result of an acquired brain injury. People who have been in automobile accidents and suffer a traumatic brain injury generally have difficulty with thinking, paying attention and behavior after the injury. Other brain injuries, like strokes, may cause cognitive impairments including difficulty speaking, or hemiplegia (paralysis of one side of the body). Thus, some individuals have both a cognitive disability and a physical disability related to the same cause [37].

Recognition and understanding of cognitive disabilities and the potential areas and skills that may be impaired is critical when planning for and implementing a disaster plan. Communication skills allow people to gain a shared understanding of what to do and what to expect during a disaster. However, people with limited communication skills have difficulty doing that and will have difficulty following directions provided to them. They may not understand public service announcements because they don't have the vocabulary or language ability to understand those. As well they may not be able to read guidance or information sent out. People with cognitive disabilities don't always have other disabilities such as physical, hearing or vision disabilities. As well, there are people without cognitive disorders that don't communicate verbally including people that are deaf or hard of hearing that might use sign language as their primary form of communication. First responders will need to consider the possibility of the presence of a cognitive disability and difficulties with communication when they come in contact with people that don't respond to interaction with them in a typical or expected manner. Because the populations of people with cognitive impairments vary, groups will be described individually to help better define how to think about planning for and working with them.

People with intellectual disabilities

Intellectual disabilities represent approximately 1% of the American population meaning that about 7 million Americans have intellectual disabilities [49]. The American Association on Intellectual and Developmental Disabilities (AAIDD) defines intellectual disability as "a disability characterized by significant limitations in both intellectual functioning and in adaptive behavior, which covers many everyday social and practical skills" and occurs before the age of 18 [50]. Intellectual functioning is measured using cognitive tests and defined by what is known as an intelligence quotient or IQ. The mean IQ of the general population is around 100 and an IQ consistent with an intellectual disability is typically below 70. Adaptive behavior includes daily living skills such as personal care, language and communication, number skills with money, time, and other number management, and social skills. Generally people with intellectual disabilities learn both intellectual or academic skills and daily living skills more slowly than those without, however, within the group of people with intellectual disabilities a wide range of abilities exists. Understanding the range of abilities within the population of people with intellectual disabilities will not only help with planning for a disaster, but also assist those first responders that may need to help a person with an intellectual disability who may have become separated from familiar people.

Intellectual disabilities (ID) are divided into four groups based roughly on IQ, although often the two groups with lower IQs are combined as functionally their support needs are similar [51]. People with mild intellectual disabilities have an IQ between 50 and 70. Many of them can hold a conversation, read and write some, and do some basic math. They often exhibit difficulties with reasoning, problem solving, planning, abstract thinking, and judgment which may worsen in times of stress such as during a disaster [52]. However, in general they understand directions and can follow them. Moderate intellectual disability is characterized by an IQ between about 35 and 50 which has a greater impact on their ability to communicate and understand communication. People with moderate ID can learn basic safety, how to travel to and from familiar places, and to complete many daily activities independently or with supervision. For the remaining two groups of individuals, their ID has a significant impact on their ability to function independently. Their abilities to understand language and communicate are often limited and they may communicate through facial expressions and gestures. Typically they need assistance with everyday tasks and supervision to remain safe in the community. The differences between these groups and their everyday needs requires a somewhat different approach to planning for and thinking about how to support them before, during and after a disaster.

People with intellectual disabilities primarily live in the community with family, alone in their own home or in small group settings with other individuals with intellectual disabilities. State programs in conjunction with the federal government fund services that promote independence, living in the community and participation in employment, volunteer work and leisure activities [53]. In the past, many individuals with intellectual disabilities lived in state-run or private institutions with little contact with the outside world. However, in the last half century or more, families, states, providers and other stakeholders have worked together to create individual support plans and homes where people with intellectual disabilities can live in the community. Backed by legislation promoting the civil rights of people with intellectual disabilities along with actions such as the launching of President John F. Kennedy's President's Panel, institutions closed and people with intellectual disabilities moved into or stayed in the community supported by a network of community services including employment, education, and opportunities to participate in their community in a meaningful way [54]. Consequently people with intellectual disabilities live in the same neighborhoods, apartment buildings, towns, cities and suburbs as everyone else in homes that do not differ from those in the neighborhood.

In addition to their cognitive disabilities, people with intellectual disabilities may have health problems both unique to the cause of their intellectual disability and those found in the general population. The three most common causes of intellectual disabilities are Down Syndrome or Trisomy 21, Fetal Alcohol Spectrum Disorder (FASD), and Fragile X Syndrome [51,55]. Sometimes intellectual disabilities with genetic causes may have associated congenital anomalies such as congenital heart disease or eye problems which is the case in both Down Syndrome and FASD. Others may have brain malformations with associated disabilities related to brain function such as cerebral palsy. Seizures disorders or epilepsy are common in people with intellectual disabilities and may be present in 20% to 30% of people with ID compared to about 1.2% of the general population [56,57]. Other developmental conditions such as Autism Spectrum Disorder (ASD) and Attention Deficit Hyperactivity Disorder (ADHD) coexist with intellectual disabilities with a frequency of 10% and 15% respectively [58]. Mental health conditions such as depression and anxiety are also commonly found in people with intellectual disabilities.

Physical health conditions common in the general population also affect people with intellectual disabilities [55]. Obesity and associated type 2 diabetes, high blood pressure and increased cholesterol present in the ID population as in the general population. Coronary artery disease, congestive heart failure and chronic obstructive pulmonary disease are also common and not only have the same treatments, but also produce the same level of functional limitation in their severe form in both populations. Those with severe heart or lung disease may take a number of medications and also be treated with oxygen. Heart disease and some other physical health conditions in the population of people with ID may be complicated by the combination of congenital conditions such as congenital heart disease and the impact of acquired disease like heart disease from high cholesterol. This is commonly seen in individuals with Down Syndrome that often have both congenital heart disease and obesity leading to acquired heart disease. This can result in more complex conditions than in the general population.

People with other cognitive disabilities (brain injury, dementia, stroke)
Cognitive disabilities are not just congenital in nature. Many adults have acquired disabilities that impact their cognition, ability to think and decision-making capability [47,48,59]. Some of these conditions also cause behavioral difficulties and associated mental illness such as anxiety and depression. Others may also be associated with physical disabilities or limitations. These disabilities are related to conditions that affect the brain, producing either diffuse or localized damage. Some conditions, such as traumatic brain injuries, are static in nature meaning that the damage was done at a point in time and the condition doesn't progress, while others are progressive in nature. Progressive conditions consist of: (1) various types of dementia or neurocognitive disorders; (2) autoimmune disorders, such as multiple sclerosis where the brain's nerves are damaged by the body's immune system; (3) genetic disorders, including Huntington's disease where the nerve cells break down over time; and (4) chronic infectious diseases such as Human Immunodeficiency Virus (HIV) and syphilis [60–66]. In addition, chronic exposure to toxins, such as alcohol, can result in damage to the nerve cells in the brain, creating cognitive difficulties and dementia.

Dementia

Multiple types of dementia exist with Alzheimer dementia, the most common form of progressive cognitive deterioration, making up 60% to 80% of dementias in the United States, and being the most familiar one [59,60]. Recently, dementia prevalence has increased dramatically in the United States and across the world, becoming 1 of the top 10 causes of death in the United States. Affecting mostly elderly people, dementia impacts not only the person, but their family because individuals with dementia require significant amounts of care as the condition progresses. Dementias progress over a variable period of time from mild to severe and affect memory, decision making ability, and daily functioning. People with early stage disease have mild symptoms affecting some areas including memory, executive functioning, attention, language and visual spatial skills. Many people at this stage do not seem different from adults without dementia as most of their cognitive functioning is still intact. However, in compensation for their decline in memory, they may confabulate, mixing present reality with past or truth with stories [61]. For example, an older woman with dementia may say that her husband will be home from the store any minute when in fact he has been dead for many years. This is important to recognize as these individuals may need more assistance during a disaster than it may appear at first. As dementia progresses the symptoms worsen, and people become less able to function independently. Periods of hallucinations and delusions with paranoia occur as do ones of confusion, which may create suspicion in the mind of the person with dementia. They may refuse to allow first responders entry to their home or refuse to evacuate in a disaster, not understanding the danger. Anxiety also plays a role in behavior in people with dementia. People in end stage dementia may become unable to move around independently or communicate in any manner and require transportation by wheelchair or stretcher.

Vascular disease. Another condition that impacts cognitive functioning is vascular disease [35–37]. This causes decreased blood flow to arteries in many parts of the body including the heart (cardiovascular disease), and the brain (cerebrovascular disease). Cerebrovascular disease leads to stroke with the potential for both cognitive loss and physical motor impairment. Long standing cerebrovascular disease and the occurrence of multiple strokes produces a vascular dementia which has an effect on how a person thinks and performs tasks that require thinking. Vascular dementia differs from Alzheimer in that the early stages are marked by reasoning and judgment difficulties, and memory difficulties generally don't become problematic until later in the course. Other cognitive difficulties may occur with strokes without the development of dementia. Strokes affecting the left side of the brain create difficulties with speech and understanding language, making communication challenging. Strokes also result in difficulty organizing and a cautious, hesitant manner of approaching activities in addition to varying degrees of paralysis of the right side of the body. Right sided strokes leave the person with left sided paralysis, but also behaviors referred to as "agnosia" (neglect) where the person ignores the left side of their body and doesn't use it in everyday activities. In addition, difficulties with spatial tasks, such as reading a map or judging the location of objects in one's surroundings, make everyday activities like climbing stairs, eating, and getting dressed difficult. Other problems related to right-sided strokes include lack of emotion, difficulty with non-verbal communication, understanding body language, gestures and humor, and the inability to focus for longer periods of time. Strokes on either side of the brain give rise to behavior and challenges with cognitive functioning that will require consideration when planning for and implementing a disaster plan.

Other progressive brain conditions

Other progressive brain conditions include those caused by inflammation, infection, and damage produced through different mechanisms impacting cognition and other brain function. Multiple sclerosis,

HIV, syphilis, Huntington's disease, and alcohol all damage nerve cells in various parts of the brain to create a recognizable pattern of functional impairments that limit daily functioning [62–64,66,67]. Similar to vascular disease, most of these diseases cause both motor and cognitive difficulties. For example, people with late stage neurosyphilis (syphilis infection that has spread to the brain and spinal cord) walk with a wide-based gait described as "staggering" or "slapping" [62]. The gait in Huntington's disease is characterized by slowed pace, difficulty initiating steps and variability in stepping patterns resulting in balance issues [63]. Multiple sclerosis (MS) also can cause a slow gait with variability in stepping pattern and likewise can result in spasticity (stiffness) that leads to shortened steps and limitations in joint movement [64]. As an episodic condition MS symptoms wax and wane over time.

In addition to gross motor issues, these conditions may impact other motor function. For example, slurred speech, slowed speech and articulation difficulties can occur in MS and make people's speech hard to understand or they may appear to be intoxicated when it is the neurologic condition causing the slurring [68]. These speech difficulties may be present after a stroke as well. Speech problems and cognitive deficits can be found together in individuals with neurologic conditions impacting their ability to communicate on a daily basis [59]. This impact on the ability to communicate affects conversation with strangers more than with family and friends who learn to understand the nuances of the person's language over time. First responders and other casual contacts do not have this advantage and may find communication difficult or may mistakenly assume that because the person's speech is affected that they are intoxicated. As well, combining this with hearing impairment common to many older adults and anxiety in the face of an emergency, intelligibility may drop even further. Keeping these issues in mind when responding to an emergent situation, will assist all involved in more effective communication.

In addition to motor limitations, these progressive neurologic conditions all are associated with varying degrees of cognitive functioning difficulties. Memory problems, difficulty in judgment and planning, and personality changes can be seen. Many conditions have more global effects on the brain and impact both right and left sided brain functions, impairing language and organizational skills as well as direction sense, memory and behavior. Some individuals lose the ability to read, write and learn new skills while others maintain those abilities. Confusion may occur and can be situational, worsening in the presence of stressful events and new circumstances. Many of these conditions can lead to loss of executive functioning including the ability to concentrate and focus, control impulses, and make daily decisions and can have a significant impact in an emergent situation.

Brain injury

Traumatic brain injury (TBI) acquired from head injuries in car or bike accidents, falls, or other causes exemplify many of both the cognitive and motor issues discussed previously [64]. Other types of acquired brain injury exist from causes such as hypoxia or lack of oxygen to the brain, infections including meningitis or encephalitis from bacteria or viruses, or lead poisoning. Strokes, discussed earlier, represent an acquired brain injury. As in strokes, the consequences of traumatic brain injury reflect the damaged areas of the brain. Common outcomes include motor difficulties with unsteady gait or inability to walk, cognitive deficits, personality changes, and increased likelihood of developing mental illness. Irritability, impulsivity, increased aggression and mood lability characterize some of the personality changes that can occur after a brain injury. Mood instability leads to exaggerated and inappropriate responses to stimuli in the environment which complicates everyday life. Difficulty with everyday tasks such as untangling a shoelace knot can lead to frustration and anger outbursts in a situation that previously would have been a minor annoyance. People with these changes from TBI often are unaware of the extent of the changes, although they recognize that they are somehow different.

Sequelae or outcomes from other acquired brain injuries differ depending on the type of injury and cause. However, many of these, including brain infections and toxins, produce a more homogenous damage across the brain. This results in cognition decreases in cognition, motor impairments, and sometimes decreased levels of consciousness including coma [69]. Post-encephalitis, people may develop seizure disorders or epilepsy, hearing impairment, anxiety and mood swings and fatigue, in addition to outcomes similar to other causes of brain injury including memory and cognitive issues, personality changes, difficulty with planning and organizing, and motor problems such as balance difficulties.

Despite these difficulties many people with brain injury or medical conditions affecting brain function live in the community with varying levels of support. Some live independently and others with family. Many have home and community-based services to help them function on a day to day basis. Some require higher levels of support and may reside in specialized environments structured to support the cognitive and behavioral issues that follow from a brain injury. Some, especially those with coma, may live in institutional settings such as nursing facilities. Regardless of where they live, success in managing emergency responses and disasters calls for understanding the underlying issues in order to support people with a brain injury.

Behavioral health conditions

Behavioral health, as defined by the Substance Abuse and Mental Health Services Administration (SAMHSA) in the context of mental health conditions, refers to promoting mental health, resilience and wellbeing; treating mental health and substance use disorders; and supporting recovery in people with these conditions, their families and communities [70]. Mental health conditions encompass affective disorders such as depression and bipolar disorder, psychotic disorders like schizophrenia and schizoaffective disorder, and anxiety disorders including panic disorders, obsessive compulsive disorders and generalized anxiety. Substance use disorders, like opiate or alcohol use disorders may accompany mental health disorders or be found separate from them. All of these disorders are prevalent in the general population. Estimates are that 26% of Americans suffer from a diagnosable mental illness in a given year [71]. Depression and anxiety are the most prevalent mental health disorders affecting about 9.5% and 18% of adults, respectively, in a year. Common patterns of co-occurrence of mental health and substance use disorders include depression with substance use and anxiety. Some episodes of mental health disorders will remit while others will become persistent and even lifelong. Serious mental illness (SMI), defined as any mental illness that causes serious functional impairment in one or more life activities, affects about 5% of the population and has a significant impact on daily living for these individuals [72]. Conditions such as schizophrenia, with a prevalence of about 1% of the population, commonly cause SMI.

Anxiety disorders make up a high proportion of the mental illness in the US population and many people have symptoms of anxiety that don't reach the threshold of the diagnostic criteria. Nonetheless, these symptoms can impact people's lives and functioning especially in times of increased stress such as a disaster. One specific anxiety disorder, post-traumatic stress disorder (PTSD), is not only prevalent, but often a consequence of experiencing previous trauma such as occurs in an emergency or disaster [73]. Many people, including disaster and relief workers that have lived through a traumatic event such as an earthquake will have feelings of being glad to be alive initially although these may be followed by fear and anxiety, difficulty sleeping and anger. For some, these feelings can continue long after the event. Persistent intrusive memories and flashbacks of the event, both unwanted and triggered by the environment, may be accompanied by behavior avoiding things that remind them of the event, negative feelings about themselves including hopelessness, continued irritability and fearfulness characterize PTSD. It is present in the general population in about 5% to 6% of people. However, particular populations of people have a higher prevalence, including military veterans with a prevalence of about 10% to 15%. Among immigrants, many of whom have experienced war, torture, and life threatening events, PTSD incidence ranges from 14% to 19% [73,74]. African Americans face higher rates of PTSD than whites in the United States with about a 1.2 times higher risk [75]. The behavioral and emotional response that people with PTSD may have to disasters is an important consideration when planning for disaster response, such as sheltering, as these types of situations may become triggers. First responders and others involved in disasters and emergencies should be cognizant of PTSD symptoms within themselves, and should engage in activities to promote self-recovery and resiliency, including post event, and should seek help early if symptoms develop.

Mental illness and people with disabilities and other health conditions

The prevalence of mental illness in people with disabilities exceeds that of the general population, according to a study using data from the Behavior Risk Factor Surveillance Survey (BRFSS). The BRFSS is a public health telephone survey used to gather health and health risk information in the US population in order to research the prevalence of various conditions in people with disabilities compared to the general population [76]. Disability was defined as requiring the use of specialized equipment such as a mobility device or having limitations in performing daily activities, which is consistent with the public health of individuals with access and functional needs discussed earlier. In this study the group of individuals with disabilities had significantly higher levels of anxiety and depression, about three times more, than those without disabilities. Another factor used to measure mental distress and the risk for mental health difficulties and adverse health is mentally unhealthy, days generally reported as the number of mentally unhealthy days in a 30 days period [77]. Frequent mental distress present on 14 or more days in a 30 days period is associated with poorer health, increased rates of mental health conditions especially depression, chronic diseases and functional limitations, and adverse health behaviors such as smoking and physical inactivity. About 33% of people with disabilities in the study experienced mental distress at this level which is 4.6 times that of their peers without disabilities. In addition, people living below the poverty level, a more common occurrence in people with disabilities than those without, reported mental distress 70% more than those in higher income groups. The overlap between mental health and disability impacts the ability of people with disabilities to respond to and recover from disasters.

Mental illness and physical health conditions are also linked as people with cancer, diabetes and heart conditions have a higher rate of depression and people with mental illness have a higher rate of physical conditions including diabetes, obesity and heart disease [78,79]. Depression is common in people with chronic health conditions especially those leading to difficulties with performing every day activities. Many of the conditions that cause disability such as stroke, Alzheimer dementia, Multiple Sclerosis result from changes to the brain and also change the ability to function, increasing stress and anxiety. This combination often leads to depression. However, depression also impacts everyday functioning and interferes with the ability to access care, exercise, eat well and other healthy behaviors. This in turn causes physiologic changes in heart rate, metabolism, stress hormones and other physical factors, compounding physical health conditions such as heart disease, diabetes, and increasing risks for stroke, lung disease, dementia and additional conditions. Studies show that people with mental illness, particularly serious mental illness, live shorter lives and have more physical health complications than the general population [79,80]. A number of factors contribute to these poorer outcomes, including: (1) decreased access to and utilization of care, especially preventive care; (2) psychotropic medication side effects; and (3) unhealthy behaviors including smoking and those leading to obesity, such as poor diet and lack of exercise. People with SMI have a shorter lifespan and more physical illness than the general population. They are also more likely to have physical access and functional needs populations and participation in Medicaid waiver programs for people with disabilities. In addition, this population tends to be less well prepared for disasters than the general public and has greater difficulty coping during and after a disaster.

Prevalence of mental illness

People with mental health disorders make up roughly a quarter of the US population. Most people live in the community, have families, work and are generally indistinguishable from the general population. People with serious mental illness by definition have more difficulty with everyday activities. Those that are most affected may live with family or alone, although stable housing is a significant issue for this population [81]. They experience higher rates of homelessness and are vastly overrepresented in jails and prisons [82]. Estimates are that 20% of the prison population have a serious mental illness, 50% to 75% have any mental illness, and 30% to 60% have substance use disorders. People with mental illness and substance use disorders without stable housing tend to live between homelessness, shelters and hospitals. A continuum of supported housing with various levels of assistance exists for people who need these levels of support [81]. However, the need exceeds the capacity in many places and these options may not be available in others. The most supportive of these programs have round the clock staff. In the lowest level of supported housing, staff may visit regularly based on the need of the person and the program, but not necessarily daily. There are also options with levels of support between those two in intensity. Housing subsidies through federal, state and local programs including federal section eight housing and housing support for people with disabilities exist to support this population.

Post-disaster mental health conditions

Some evidence exists that disasters can result in the development of some types of mental illness in survivors [80]. With the disruption in routine and displacement from home, people with existing SMI may exhibit more symptoms of their illness. In general people with SMI tend to have lower levels of self-esteem and social support, with the lowest levels found in individuals with schizophrenia. Many lack or have weak coping skills, including the use of approach coping strategies that people use to manage stressful events or situations by actively focusing on the event or situation. This problem-focused coping mechanism employs techniques and strategies such as logical analysis of the situation, positive reappraisal of the stressor, problem solving, and seeking guidance and support from others and is commonly used to reduce stress. People with schizophrenia and bipolar disorder have been found to use an avoidance response more frequently, which is an indicator of poorer mental health. Post-disaster this can lead to persistence of the stress symptoms. Evidence shows that the prevalence of schizophrenia and related psychotic disorders and bipolar disorders does not increase after a disaster. However, post-disaster major depression and PTSD have been observed, with PTSD being the most common post-disaster mental illness to develop. In one study of 1993 Midwestern flood survivors noted that at 4 months after the floods 20% of the survivors were experiencing major depression which decreased to 11% at 1 year post flood. Similarly, the Oklahoma bombings in 1995 resulted in a 38% increase in new major depression diagnoses with all but about 10% in remission 7 years later. Impacts of disasters on mental illness can be serious and lasting, but are not always persistent.

Autism spectrum disorder

One additional condition that is important to include in this discussion illustrates the co-existence of many of the issues previously discussed. People with Autism Spectrum Disorder (ASD), a neurodevelopmental disorder, exhibit difficulty with social interaction, communication including non-verbal communication, and usually have a restricted repertoire of interests with repetitive behaviors [83]. ASD prevalence varies, but estimates put it at 18.5 per 1000 children in the sampled states across the United States. It is associated with cognitive and learning disabilities with some people with ASD also having an intellectual disability. Mental health disorders commonly co-occur with ASD contributing to the difficulties that individuals may have functioning day to day [84]. In a clinic-based study of people with ASD, 28% were diagnosed with attention-deficit hyperactivity disorder, 20% with an anxiety disorders, 11% with depression, 5% with bipolar disorder and 4% with a disorder on the schizophrenia spectrum. These frequencies exceed those in both population-based studies of people with ASD and the presence of these conditions in the general public. People with ASD generally have difficulty with change and also have difficulty communicating, and sometimes understanding communication. These difficulties may be present even in the absence of an intellectual disability and combined with attention problems, anxiety and other mental health conditions may create a scenario for destabilization during a disaster where everyday life is disrupted. Communication is important to maintain safety and promote good decision-making. While many people with ASD live with family or others who know them and understand how to communicate with and support them, recognizing the potential responses of a person with ASD can assist with finding the right supports for them through and after a disaster.

Children with disabilities

Up to this point, the population focus has been on adults. However, many children and adolescents experience chronic medical conditions, developmental disorders and mental health conditions that impact their cognitive and physical functioning, causing functional limitations compared to their peers. The CDC states that about 25% of children ages 2 to 8 years have a chronic physical or mental health condition [85]. Many of these conditions present in childhood and persist into adulthood. The ASD prevalence studies primarily involve school-aged children, and children with ASD have the same co-morbidities or co-occurring mental health and cognitive conditions (e.g., anxiety and learning problems) that adults do. Estimates show that about 7.4% children aged 3 to 17 years have a behavior problem and 9.4% of children aged 2 to 17 years carry a diagnosis of Attention Deficit Hyperactivity Disorder (ADHD) [86]. About 2.3% and 7.1% of children have depression and anxiety, respectively. Anxiety commonly occurs with depression in children as 73.8% of children with depression also have anxiety. Fewer children have Serious Mental Illness (SMI) such as schizophrenia or bipolar disorder since those conditions typically do not appear until late adolescence or early adulthood. Nonetheless, childhood mental illness is both relatively common and significantly impacts children's day to day functioning.

In addition to behavioral health conditions, children and adolescents also have physical health and developmental conditions that lead to functional impairments and the use of medical technology. Asthma, allergies and obesity are among the most frequent conditions in children. Asthma tops the list as the most prevalent condition and significantly impacts school attendance, participation in activities, and daily functioning. It is most common in minority children, those living in urban areas, and those who are poor [85]. While not common, conditions like congenital heart disease, cystic fibrosis, and type 1 diabetes also affect the health and functioning of children, requiring daily treatment with medication, for example nebulizers to help breathing, and diet needs that might differ somewhat from the typically developing child. Developmental disorders such as cerebral palsy, spina bifida, and muscular dystrophy impact children's abilities to move and sometimes affect cognition as well. Physical health, mental health, and developmental conditions frequently co-occur and children with these conditions exhibit access and functional needs that can impact their ability to participate in the community without support. Their physical needs do not differ greatly in principle from those of adults with cognitive, physical and mental health conditions, and include access and functional needs, making it possible to understand their needs and include them in planning. It is also important to consider a child's developmental age.

Some children with disabilities may have a lower developmental than chronological age and some children may regress due to a disaster or emergency. All children, including those with disabilities, differ from adults in that developmentally they are not mature and not expected to be mature. However, consideration needs to be made for how to approach a child with regard to how much they talk, what they can understand at their developmental level, and how to support them during a disaster. Studies of children's responses to disaster show that their reactions differ from adults, and that they may be impacted more by the perceived threat than the actual threat [87]. Children experience disasters in the context of their family and community, and their ability to cope with the disaster is often a function of their parents' ability to support them through this. Post disaster children often face fears and anxiety, difficulty sleeping, problems concentrating, disobedient behaviors, and depression. All of these may also be felt by adults, however, children have less well developed coping strategies and ways to manage the stress and feelings of anxiety as well as less ability to express them verbally. Thus, planning to support children during and after a disaster is important to assure their ability to successfully weather the stresses and fears without long standing sequelae.

Coping during disasters

Recognizing the co-occurrence of conditions helps identify issues that may be faced by the different populations of children and adults with access and functional needs. In addition to physical support, behavioral and emotional support may also be needed in a time of crisis. While all people need these supports during a disaster, people with access and functional needs may be less prepared and cope more poorly in a stressful situation. These populations have fewer resources, more tenuous access to adequate food and shelter, and are more dependent on public transportation and services. The disruptions to everyday life's routines during a disaster, with unavailability of public transportation, food and drug store closures, and high levels of media coverage impact physical day to day functioning and also create an environment of anxiety and stress. This can magnify an existing level of anxiety, leading to ineffective responses in a disaster. Understanding and recognizing what behaviors might be encountered can lead to providing better support both during and after a disaster. Knowing that many people with access and functional needs may be less well prepared for a disaster, working to include them in local and state planning as well as encouraging individual self-preparedness plans is critical to help them be prepared for a disaster.

In recognition of the needs of these populations, the Centers for Medicare and Medicaid Services (CMS) promulgated a disaster planning rule for multiple types of providers including hospitals, community health centers, and other providers that serve people in the community [88]. Additionally, many disability and mental health state program regulations include written requirements for disaster plans to guide service providers to create disaster plans including emergency evacuation plans for their service recipients. Resources from the Federal Emergency Management Agency (FEMA) and state and local emergency management agencies can help with the continuum of planning needs from preparedness to disaster response to the aftermath. Some of these resources specifically address access and functional needs issues, either by population or in general. In reviewing the characteristics and needs of the various populations with access and functional needs, it has been demonstrated that many of their needs may be similar in nature and planning from an all-disability construct will serve to include and meet the needs of these populations.

Questions for thought

(1) Describe the access and functional needs issues someone might have related to evacuating their home during a disaster or emergency. Ask someone you know or visit the website of one of the groups discussed in this chapter to seek this information.

(2) Describe the access and functional needs issues someone might have related to understanding a disaster evacuation alert and following evacuation instructions. Ask someone you know or visit the website of one of the groups discussed in this chapter to seek this information.

(3) Choose a factor leading to health inequities and discuss how it impacts one of the populations of people with access and functional needs described in the chapter.

References

[1] Oxford Languages, Oxford English Dictionary, Oxford University Press, https://www.google.com/search?q=what+is+language&rlz=1C1GCEU_enUS821US822&oq=what+is+language&aqs=chrome..69i57.2151j0j7&sourceid=chrome&ie=UTF-8.
[2] National Center on Disability and Journalism, Disability language style guide, Walter Cronkite School of Journalism and Mass Communication Arizona State University, https://ncdj.org/style-guide/.
[3] Centers for Disease Control and Prevention. Resources & style guides for framing health equity & avoiding stigmatizing language, inclusive communication principles, 2021. December 16 https://www.cdc.gov/healthcommunication/Resources.html.
[4] US Equal Employment Opportunity Commission. The rehabilitation act of 1973: Sections 501 and 505, https://www.eeoc.gov/statutes/rehabilitation-act-1973.
[5] Crocker AF, Smith SN. Person-first language: are we practicing what we preach? J Multidiscip Healthc 2019;12:125–9. https://doi.org/10.2147/JMDH.S140067.
[6] The ARC. The power of words: People first language, https://www.thearc.org/who-we-are/media-center/people-first-language.
[7] Integrated Living Opportunities, Organizations, https://www.ilonow.org/resources/organizations/.
[8] Ladau E. Why person first language doesn't always put the person first. MCIE Think Inclusive Blog 2021;(Jul 7). https://www.thinkinclusive.us/why-person-first-language-doesnt-always-put-the-person-first/.

[9] Streeter LE. The continuing Saga of people-first language. Braille Monitor, National Federation of the Blind; May 2010. https://nfb.org//sites/default/files/images/nfb/publications/bm/bm10/bm1005/bm100509.htm.

[10] National Association of the Deaf. Community and culture—Frequently asked questions, 2014. Retrieved 2014-05-22 https://www.nad.org/resources/american-sign-language/community-and-culture-frequently-asked-questions/.

[11] Sinclair, J. Why I dislike person-first language, Autism Mythbusters Retrieved 7 January 2016.

[12] Brown L. Identity-first language, autistic self-advocacy. Network 2020. https://autisticadvocacy.org/about-asan/identity-first-language/.

[13] Debbaudt, D., Information for first responders and other professionals, autism safety project, Autism Speaks, https://www.autismspeaks.org/information-law-enforcement.

[14] U.S. Department of Health and Human Services, Office of Assistance Secretary for Preparedness and Response. Federal emergency management agency, HHS/ASPR access and functional needs fact sheet., September 08, 2020, https://www.phe.gov/Preparedness/planning/abc/Pages/AFN-FactSheet.aspx.

[15] National Center on Birth Defects and Developmental Disabilities, Centers for Disease Control and Prevention. Impairments, Activity Limitations, and Participation Restrictions. Disability and Health Promotion; September 16, 2020. https://www.cdc.gov/ncbddd/disabilityandhealth/disability.html.

[16] Kailes JI, Enders A. Moving beyond "special needs"—A function-based framework for emergency management and planning. J Disabil Policy Stud 2007;17(4):230–7.

[17] United States Department of Health and Human Services, Office of the Assistant Secretary for Preparedness and Response. Public preparedness fact sheet, at risk individuals, https://www.phe.gov/Preparedness/planning/abc/Pages/atrisk.aspx.

[18] United States Department of Health and Human Services, Office of the Assistant Secretary for Preparedness and Response. Public preparedness fact sheet, access and functional needs, https://www.phe.gov/Preparedness/planning/abc/Pages/afn-guidance.aspx.

[19] The ASL app. FAQ on Deaf Culture! The ASL app. HELLO@THEASLAPP.COM, INK & SALT LLC, 2019. Retrieved from the world wide web April 3 http://theaslapp.com/faq.

[20] World Health Organization. Mental Health and emergencies fact list Mental health in emergencies, https://www.who.int/news-room/fact-sheets/detail/mental-health-in-emergencies.

[21] US Department of Health and Human Services, Office of Assistance Secretary for Preparedness and Response. Federal emergency management agency, FEMA's functional needs support services guidance., September 08, 2020, http://www.phe.gov/preparedness/planning/abc/Pages/functional-needs.aspx.

[22] Rickert T. Independent living, mobility for all, https://www.independentliving.org/mobility/mobility.pdf.

[23] Litman T. Evaluating accessibility for transport planning: measuring people's ability to reach desired goods and activities., March 18, 2019, http://www.vtpi.org/access.pdf.

[24] The Washington Group on Disability Statistics. The Washington group short set on functioning (WG-SS)., April 29, 2020, https://www.washingtongroup-disability.com/fileadmin/uploads/wg/Documents/Questions/Washington_Group_Implementation_Document__2_-_The_Washington_Group_Short_Set_on_Functioning__1_.pdf.

[25] Center for Disease Control and Prevention. Prevalence of disability and disability type among adults, United States., 2013, https://www.cdc.gov/ncbddd/disabilityandhealth/features/key-findings-community-prevalence.html.

[26] Stevens AC, Carroll DD, Courtney-Long EA, Zhang QC, Sloan ML, Griffin-Blake S, Peacock G. Adults with one or more functional disabilities—United States, 2011–2014. Morb Mortal Wkly Rep 2016;65(38):1021–5. https://www.cdc.gov/mmwr/volumes/65/wr/mm6538a1.htm.

[27] Wiese D, Stroup AM, Crosbie A, Lynch SM, Henry KA. The impact of neighborhood economic and racial inequalities on the spatial variation of breast cancer survival in New Jersey. Cancer Epidemiol Biomark Prev 2019. https://doi.org/10.1158/1055-9965.EPI-19-0416. https://www.ncbi.nlm.nih.gov/pubmed/31649136.

[28] Anon. Poverty's association with poor health outcomes and health disparities. Health Aff 2014;(October 30). https://doi.org/10.1377/hblog20141030.041986.

[29] Centers for Disease Control and Prevention. Proportion of Americans with which kinds of disabilities, Infographic, https://www.cdc.gov/ncbddd/disabilityandhealth/infographic-disability-impacts-all.html.

[30] Hamel L, DiJulio B, Firth J, Brodie M. Kaiser health tracking poll: November 2014., November 21, 2014, https://www.kff.org/health-reform/poll-finding/kaiser-health-tracking-poll-november-2014/.

[31] Bialik K. 7 facts about Americans with disabilities., July 27, 2017, https://www.pewresearch.org/fact-tank/2017/07/27/7-facts-about-americans-with-disabilities/.

[32] Olmstead Supreme Court Decision, https://www.hhs.gov/sites/default/files/olmstead_decision.pdf.

[33] United States Senate, Health, Education, Labor, and Pensions Committee, Harkin T. Separate and unequal: States fail to fulfill the community living promise of the Americans with disabilities act., July 18, 2013, https://www.help.senate.gov/imo/media/doc/Olmstead%20Report%20July%202013l.pdf.

[34] Manini T. Development of physical disability in older adults. Curr Aging Sci 2011;4(3):184–91. https://www.ncbi.nlm.nih.gov/pmc/articles/PMC3868456/.

[35] Ahto M, Isoaho R, Puolijoki H, Laippala P, Romo M, Kivelä SL. Functional abilities of elderly coronary heart disease patients. Aging (Milano) 1998;10(2):127–36. https://www.ncbi.nlm.nih.gov/pubmed/9666193.

[36] Benjamin EJ, Blaha MJ, Chiuve SE, Cushman M, Das SR, Deo R, de Ferranti SD, Floyd J, Fornage M, Gillespie C, Isasi CR, Jiménez MC, Jordan LC, Judd SE, Lackland D, Lichtman JH, Lisabeth L, Liu S, Longenecker CT, Mackey RH, Matsushita K, Mozaffarian D, Mussolino ME, Nasir K, Neumar RW, Palaniappan L, Pandey DK, Thiagarajan RR, Reeves MJ, Ritchey M, Rodriguez CJ, Roth GA, Rosamond WD, Sasson C, Towfighi A, Tsao CW, Turner MB, Virani SS, Voeks JH, Willey JZ, Wilkins JT, Wu JH, Alger HM, Wong SS, Muntner P. Heart disease and stroke statistics-2017 update: a report from the American Heart Association. Circulation 2017;135(10):e146–603. https://doi.org/10.1161/CIR.0000000000000485. Epub 2017 Jan 25 https://www.ncbi.nlm.nih.gov/pubmed/28122885.

[37] Armour BS, Courtney-Long EA, Fox MH, Fredine H, Cahill A. Prevalence and causes of paralysis—United States, 2013. Am J Public Health 2016;106(10):1855–7. https://doi.org/10.2105/AJPH.2016.303270. https://www.ncbi.nlm.nih.gov/pmc/articles/PMC5024361/.

[38] Kraus L, Lauer E, Coleman R, Houtenville A. 2017 disability statistics annual report. Durham, NH: University of New Hampshire; 2018.

[39] Anon. 2017 Disability statistics annual report, https://disabilitycompendium.org/sites/default/files/user-uploads/2017_AnnualReport_2017_FINAL.pdf.

[40] 2016 Disability compendium. https://disabilitycompendium.org/compendium/annual-statistics-2016.

[41] Souza MT. 2009 Worst case housing needs of people with disabilities: Supplemental Findings of the Worst Case Housing Needs 2009, March 2011. Report to Congress. U.S. Department of Housing and Urban Development Office of Policy Development and Research https://poseidon01.ssrn.com/delivery.php?ID=923094092020024086122008102004029081113000025039022062022075075122008106009003101075053052033000011045114073066006116121089089041072061038014011073081068099080099086092045092080099106009114029111122206408400612102302910310602408012610011808700705020&EXT=pdf.

[42] Dawkins C, Miller M. The picture of disability and designated housing. U.S. Department of Housing and Urban Development Office of Policy Development and Research; March 6, 2015. https://www.huduser.gov/portal/sites/default/files/pdf/mdrt_disability_designated_housing.pdf.

[43] US Department of Housing and Urban Development, Office of Policy Development and Research. 2009 worst case housing needs of people with disabilities: Supplemental findings of 2009 worst case housing needs of people with disabilities report to congress; 2011.

[44] Centers for Disease Control and Prevention. Disability and obesity, disability and health promotion., September 6, 2019, https://www.cdc.gov/ncbddd/disabilityandhealth/obesity.html.

[45] Centers for Disease Control and Prevention. Cigarette smoking among adults with disabilities, disability and health promotion., November 19, 2020, https://www.cdc.gov/ncbddd/disabilityandhealth/smoking-in-adults.html.

[46] Lexico. Oxford Dictionary, US dictionary, https://www.lexico.com/en/definition/cognition. [Accessed 17 March 2020].

[47] Miller EK, Wallis JD. Executive function and higher-order cognition: definition and neural substrates. In: Encyclopedia of neuroscience, vol. 4; 2009. p. 99–104.

[48] US Department of Health and Human Services, Centers for Disease Control and Prevention. Cognitive impairments., February 2011, https://www.cdc.gov/aging/pdf/cognitive_impairment/cogimp_poilicy_final.pdf.

[49] Maulik PK, Mascarenhas MN, Mathers CD, Dua T, Corrigendum SS. Prevalence of intellectual disability: a meta-analysis of population-based studies. Res Dev Disabil 2011;32(2):419–36. Res. Dev. Disabil. 2013;34(2):729.

[50] American Association on Intellectual and Developmental Disabilities (AAIDD). Definition of intellectual disability, https://www.aaidd.org/intellectual-disability/definition. [Accessed 17 March 2020].

[51] Committee to Evaluate the Supplemental Security Income Disability Program for Children with Mental Disorders, Board on the Health of Select Populations, Board on Children, Youth, and Families, Institute of Medicine, Division of Behavioral and Social Sciences and Education, The National Academies of Sciences, Engineering, and Medicine. 9, Clinical characteristics of intellectual disabilities. In: Boat TF, Wu JT, editors. Mental disorders and disabilities among low-income children. Washington, DC: National Academies Press (US); 2015 Oct 28. Available from: https://www.ncbi.nlm.nih.gov/books/NBK332877/.

[52] Ballan M, Sormanti M. Trauma, grief and the social model: practice guidelines for working with adults with intellectual disabilities in the wake of disasters. Rev Disabil Stud 2014;2(3).

[53] Centers for Medicare and Medicaid Services. Home & community based services, https://www.medicaid.gov/medicaid/home-community-based-services/index.html.

[54] Minnesota Department of Administration, Governor's Council on Developmental Disabilities. The reawakening 1950–1980, parallels in time a history of developmental disabilities., 2021, https://mn.gov/mnddc/parallels/five/5a/1.html.

[55] National Disability Navigator Resource Collaborative. Population specific fact sheet—Intellectual disability, fact sheets. American Association on Health and Disability; 2022. https://nationaldisabilitynavigator.org/ndnrc-materials/fact-sheets/population-specific-fact-sheet-intellectual-disability/.

[56] Understanding Intellectual Disability and Health, University of Hertfordshire. Epilepsy, http://www.intellectualdisability.info/physical-health/articles/epilepsy.

[57] National Center for Chronic Disease Prevention and Health Promotion, Division of Population Health, Centers for Disease Control and Prevention. Epilepsy data and statistics., September 30, 2020, https://www.cdc.gov/epilepsy/data/index.html.

[58] Gillberg C, Gillberg IC, Anckarsäter H, Råstam M. Overlap between ADHD and autism spectrum disorder in adults. In: Buitelaar JK, Kan CC, Asherson P, editors. ADHD in adults: Characterization, diagnosis, and treatment. Cambridge University Press; 04 April 2011. https://www.researchgate.net/publication/286356209_ADHD_in_adults_with_intellectual_disabilities.

[59] Beukelman DR, Fager S, Ball L, Dietz A. AAC for adults with acquired neurological conditions: a review. Augment Altern Commun 2007;23(3):230–42. https://doi.org/10.1080/07434610701553668. 17701742.

[60] Alzheimer Association Report. 2020 Alzheimer's disease facts and figures. Alzheimers Dement 2020;16(3):391–460. https://doi.org/10.1002/alz.12068. https://alz-journals.onlinelibrary.wiley.com/doi/full/10.1002/alz.12068.

[61] Tallberg I-M. Confabulation in dementia: constantly compensating memory systems. Neuropsychoanalysis 2007;9(1):5–17. https://doi.org/10.1080/15294145.2007.10773535. https://www.tandfonline.com/doi/abs/10.1080/15294145.2007.10773535.

[62] Bhandari J, Thada PK, Ratzan RM. Tabes dorsalis. StatPearls; June 7, 2021. https://www.ncbi.nlm.nih.gov/books/NBK557891/.

[63] Danoudis M, Iansek R. Gait in Huntington's disease and the stride length-cadence relationship. BMC Neurol 2014;14:161. https://doi.org/10.1186/s12883-014-0161-8.

[64] Cameron MH, Wagner JM. Gait abnormalities in multiple sclerosis: pathogenesis, evaluation, and advances in treatment. Curr Neurol Neurosci Rep 2011;11(5):507–15. https://doi.org/10.1007/s11910-011-0214-y. 21779953.

[65] McAllister TW. Neurobehavioral sequelae of traumatic brain injury: evaluation and management. World Psychiatry 2008;7(1):3–10. https://doi.org/10.1002/j.2051-5545.2008.tb00139.x.

[66] Winston A, Spudich S. Cognitive disorders in people living with HIV. Lancet 2020;7(7):E504–13. https://doi.org/10.1016/S2352-3018(20)30107-7.

[67] Zahr N, Kaufman K, Harper C. Clinical and pathological features of alcohol-related brain damage. Nat Rev Neurol 2011;7:284–94. https://doi.org/10.1038/nrneurol.2011.42.

[68] Rodgers JD, Tjaden K, Feenaughty L, Weinstock-Guttman B, Benedict RH. Influence of cognitive function on speech and articulation rate in multiple sclerosis. J Int Neuropsychol Soc 2013;19(2):173–80. https://doi.org/10.1017/S1355617712001166.

[69] Hansen MA, Samannodi MS, Castelblanco RL, Hasbun R. Clinical epidemiology, risk factors, and outcomes of encephalitis in older adults. Clin Infect Dis 2020;70(11):2377–85. https://doi.org/10.1093/cid/ciz635. 31294449.

[70] Substance Abuse and Mental Health Services Administration (SAMHSA), SAMHSA—Behavioral health integration, 1–3. https://www.samhsa.gov/sites/default/files/samhsa-behavioral-health-integration.pdf.

[71] Health, The Johns Hopkins University, The Johns Hopkins Hospital, and Johns Hopkins Health System. Mental health disorder statistics., 2022, https://www.hopkinsmedicine.org/health/wellness-and-prevention/mental-health-disorder-statistics.

[72] National Institute of Mental Health (NIMH). Office of science policy, planning, and communications, mental health statistics, mental health information, https://www.nimh.nih.gov/health/statistics/mental-illness.shtml#:~:text=In%202019%2C%20there%20were%20an%20estimated%2013.1%20million%20adults%20aged,%25)%20than%20males%20(3.9%25).

[73] National Center for PTSD. How common is PTSD in adults? U.S. Department of Veterans Affairs; September 10, 2021. https://www.ptsd.va.gov/understand/common/common_adults.asp.

[74] Perreira KM, Ornelas I. Painful passages: traumatic experiences and post-traumatic stress among immigrant Latino adolescents and their primary caregivers. Int Migr Rev 2013;47(4). https://doi.org/10.1111/imre.12050.

[75] Pérez Benítez CI, Sibrava NJ, Kohn-Wood L, et al. Posttraumatic stress disorder in African Americans: a two year follow-up study. Psychiatry Res 2014;220(1–2):376–83. https://doi.org/10.1016/j.psychres.2014.07.020.

[76] Kinne S, Patrick DL, Doyle DL. Prevalence of secondary conditions among people with disabilities. Am J Public Health 2004. https://ajph.aphapublications.org/doi/full/10.2105/AJPH.94.3.443.

[77] Cree RA, Okoro CA, Zack MM, Carbone E. Frequent mental distress among adults, by disability status, disability type, and selected characteristics—United States, 2018. MMWR Morb Mortal Wkly Rep 2020;69:1238–43. https://doi.org/10.15585/mmwr.mm6936a2.

[78] National Institute of Mental Health. Chronic illness and mental health: recognizing and treating depression. Mental Health Information; 2021. https://www.nimh.nih.gov/health/publications/chronic-illness-mental-health.

[79] Hert MDE, Correll CU, Bobes J, et al. Physical illness in patients with severe mental disorders. I. Prevalence, impact of medications and disparities in health care. World Psychiatry 2011;10(1):52–77. https://doi.org/10.1002/j.2051-5545.2011.tb00014.x.

[80] Substance Abuse and Mental Health Services Administration (SAMHSA). Disasters and people with serious mental illness, disaster technical assistance center supplemental research bulletin., August 2019, https://www.samhsa.gov/sites/default/files/disasters-people-with-serious-mental-illness.pdf.

[81] National Alliance on Mental Illness (NAMI). Finding stable housing., 2022, https://www.nami.org/Your-Journey/Individuals-with-Mental-Illness/Finding-Stable-Housing.

[82] Aufderheide D. Mental illness in America's jails and prisons: toward a public safety/public health model. Health Aff 2014;(April 1). https://doi.org/10.1377/forefront.20140401.038180. https://www.healthaffairs.org/do/10.1377/hblog20140401.038180/full/.

[83] Maenner MJ, Shaw KA, Baio J, et al. Prevalence of autism spectrum disorder among children aged 8 years—Autism and developmental disabilities monitoring network, 11 sites, United States, 2016 [published correction appears in MMWR Morb Mortal Wkly Rep. 2020 Apr 24; 69(16):503]. MMWR Surveill Summ 2020;69(4):1–12. https://doi.org/10.15585/mmwr.ss6904a1.

[84] Lai M-C, Kassee C, Besney R, Bonato S, Hull L, Mandy W, Szatmari P, Ameis SH. Prevalence of co-occurring mental health diagnoses in the autism population: a systematic review and meta-analysis. Lancet Psychiatry 2019;6(10):819–29. https://doi.org/10.1016/S2215-0366(19)30289-5.

[85] Division of Population Health, National Center for Chronic Disease Prevention and Health Promotion, Centers for Disease Control and Prevention. Managing chronic health conditions. CDC Healthy Schools; October 20, 2021. https://www.cdc.gov/healthyschools/chronicconditions.htm.

[86] National Center on Birth Defects and Developmental Disabilities, Centers for Disease Control and Prevention, Centers for Disease Control and Prevention. Data and statistics on children's mental health., March 22, 2021, https://www.cdc.gov/childrensmentalhealth/data.html#:~:text=7.4%25%20of%20children%20aged%20 3,have%20a%20diagnosed%20behavior%20problem.&text=7.1%25%20of%20children%20aged%20 3,4.4%20million)%20have%20diagnosed%20anxiety.&text=3.2%25%20of%20children%20aged%20 3,1.9%20million)%20have%20diagnosed%20depression.

[87] Mercuri A, Angelique HL. Children's responses to natural, technological, and Na-tech disasters. Community Ment Health J 2004;40:167–75. https://doi.org/10.1023/B:COMH.0000022735.38750.f2.

[88] Centers for Medicare and Medicaid Services. Emergency planning rule., December 1, 2021, https://www.cdc.gov/ncbddd/disabilityandhealth/infographic-disability-impacts-all.html.

The world approach to disability-inclusive disaster management

3

Jill Morrow-Gorton

University of Massachusetts Chan Medical School, Worcester, MA, United States

Learning objectives

1. Describe the role of the United Nations and its agencies in the approach to providing direction for and supporting nations in developing disability-inclusive disaster management.
2. Explain the principles set out in the Hyogo and Sendai Frameworks for inclusion of people with disabilities in disaster risk reduction.
3. Demonstrate how the public health approach to improving health supports building resilience in people with disabilities and the communities in which they live.

Integrating Mental Health and Disability Into Public Health Disaster Preparedness and Response
https://doi.org/10.1016/B978-0-12-814009-3.00008-8

55

Introduction

Disaster management and public health organizations across the world face disasters and the need to address planning, response, recovery and mitigation for their citizens. The natures of both natural and human-made disasters vary by country across the globe. International bodies, including the United Nations (UN) and World Health Organization (WHO), focus on disaster management activities and the people impacted by disasters. They have concentrated their efforts on working with people with disabilities in order to better address their needs and rights around such events. Recognition of the disproportionate social, health and economic impacts of disasters on people with disabilities has led to the development of strategies and policies to reduce these effects. Simultaneously, a growing awareness of the needs and rights of people with disabilities has helped drive some of the underlying solutions that may help strengthen disaster resilience within this population. The UN's guidance approaches disability and disaster from an emergency management standpoint while the WHO, a public health entity, focuses more on population health. Together these two viewpoints strengthen the goal of including people with disabilities as valuable and equal members of their communities.

The United Nations

In 1945 after the end of World War II, 51 countries banded together to found the United Nations (UN) with the purpose of building global peace and friendly relationships, working together to promote human rights and freedoms, and improving the lives of people living in poverty by addressing hunger, illiteracy, and disease [1]. Those goals remain today and the UN has developed a vision of having no one left behind by emphasizing the importance of inclusion, human rights, and the empowerment of all people, including those with disabilities [2]. Part of the UN's mission involves acting as spokesperson and conscience for the countries that it represents regarding global, regional and local disasters, including natural and human-caused disasters and armed conflicts. It addresses the challenges of endorsing and encouraging peace, sustainability, and social progress along with working towards strengthening poor, vulnerable populations. To this end, the UN developed a disability-inclusive stance for its disaster work, to be located within the Department of Economic and Social Affairs Office for Disaster Risk Reduction (UNDRR) [3,4].

Historically the UN recognizes the importance of disaster risk reduction and the effect that disasters of all kinds have had on the world's populations. In response to concerns about the increased number and impact of disasters on the people of the world, and especially the monetary and health including mortality costs to the world's vulnerable people, the UN developed a governmental structure to oversee their activities to address these concerns. The UNDRR evolved from a series of earlier UN governmental agencies' efforts to tackle these issues and develop frameworks and strategic plans to help address concerns about disaster management. In 1989 the UN General Assembly established the decade of 1990 to 1999 as the International Decade for Natural Disaster Reduction (IDNDR) and formed a Secretariat to oversee the disaster risk reduction work [5]. The General Assembly tasked the IDNDR with identifying and developing guidelines to apply current science and technology to disaster risk reduction with the intent that these guidelines be used to help governments build their disaster risk reduction capacity in order to minimize the impact of disasters on their population. Some of the science and technologies identified for guidelines included improved radar detection for tornados and earthquake resistant building design. The United States Centers for Disease Control and Prevention (CDC) noted that while many of these guidelines were for non-health related science and, therefore, not particularly related to public health, the impact of the guidelines on the population mortality and injury caused by these events was related [6]. In particular, over two decades in the United States, the improved ability to detect tornados has led to the development of better tornado safety guidelines, resulting in decreased deaths caused by tornados. Epidemiologic studies verified this and showed that many of these non-health related interventions decreased population morbidity and mortality, thus impacting both disaster management and public health.

In 1999, at the end of the IDNDR, the UN replaced the Secretariat of the International Decade for Natural Disaster Reduction with a new agency, the UN International Strategy for Disaster Reduction Secretariat (UNISDR), to continue the work of global disaster risk reduction [7]. The UN General Assembly made the UNISDR responsible for the coordination of disaster risk reduction activities across the UN organization including global and regional activities [2]. Through the development of an online system, UNISDR helped countries monitor and assess their progress towards disaster risk reduction. As the focal point for disaster risk reduction activities, the UNISDR was tasked with linking activities addressing a number of important related areas including economic, social, humanitarian and sustainable development interventions and promoting the integration of policies to support them. In 2019, the UN Office of Disaster Risk Reduction changed its acronym from UNISDR to UNDRR to better reflect the Office's name in English. UNDRR remains the entity responsible for translating the disaster risk reduction concept and its supporting strategic framework into practice and implementing the strategic plan [8].

Disaster reduction frameworks

To this end, a series of world meetings held over time have set down principles for world disaster risk reduction, with an increasing focus on the inclusion of people with disabilities. The initial framework, the Hyogo Framework for Action, was crafted from discussions at the World Conference for Disaster Reduction in Hyogo, Japan in 2005 [9]. This work followed from the 1994 Yokohama Strategy for a Safer World: Guidelines for Natural Disaster Prevention, Preparedness and Mitigation and its Plan of Action ("Yokohama Strategy") and took advantage of the identified gaps and lessons learned from application of the Yokohama strategy to move the field forward. In particular, the Yokohama Strategy recognized the disproportionate impact of disasters on poor populations and the fragility of urban development increasing vulnerability to natural disasters. One potential solution identified to address these issues entails encouraging the involvement of all people in their local disaster risk reduction activities. Building on the Yokohama Strategy, the Hyogo Framework for Action identified five priorities for action including: (1) making disaster risk reduction a priority; (2) having nations know their risks and take action; (3) building understanding and awareness; (4) reducing risk; and (5) being prepared and ready to act. The most recent UNISDR strategic framework for 2016 to 2021 builds on the previous work and uses the principles and strategies developed at the Third UN World Conference held in Sendai, Japan in March 2015 [10]. This instrument, the Sendai Framework, represents the goals for disaster risk reduction in the 15 years from 2015 and 2030. It expanded the concepts of the previous conferences and established four priorities for action, found in Fig. 1. These priorities reflect both the importance of disaster risk reduction and building community

1. Understanding disaster risk
2. Strengthening governance to manage disaster risk
3. Investing in resilience through disaster risk reduction
4. Enhancing disaster preparedness and use of the concept, "Build Back Better" in recovery, rehabilitation and reconstruction.

FIG. 1

Priorities of the Sendai Framework.

Excerpted from the Sendai Framework. Sendai framework for disaster risk reduction; 2015. Available from: https://www.unisdr.org/ files/43291_sendaiframeworkfordrren.pdf.

resilience, and the idea that incorporating the principles of technology and science to reduce disaster risk can also improve the life of communities.

The UNISDR bases its work on the theory that disaster risks specific to an area must be reduced in order to have sustainable development. This often involves changing the way communities and nations approach development as it requires the integration of disaster risk reduction into development. Because this requires changes in how communities and nations act, the UNISDR uses the Theory of Change (TOC) concept to develop and implement its strategic framework to guide how disaster risk reduction is incorporated into development [11,12]. The TOC method originated from attempts to evaluate the results or changes derived from community-based interventions. This methodology involves identifying the cause of the problem and the expected outcome, then applying an intervention that addresses that cause and results in the desired outcome. The key to applying the theory of change is that the potential interventions require an evidence base to show that they could effectively solve the problem at hand [13]. In this case, where the outcome is sustainable development, the disaster risks of the geographic area inform how development occurs to minimize the risks that the area will be destroyed when disaster strikes.

The United Nations Multi-Country Development Framework (UN MSDF) approach to the Caribbean illustrates how the UN uses TOC [13]. This geographic area is made up of multiple countries with varying histories. The region suffers from poor economic growth which has led to increased poverty, joblessness especially among young people, and an upsurge in crime including violent crime. These factors have contributed to poorer health as shown by a rise in adolescent pregnancies. In addition, the region is prone to hurricanes which cause significant destruction and put even greater stress on the economy and health of the population. The UN MSDF articulated four program goals to help address the needs of the Caribbean. These include sustainable economic development, healthy people, safe communities, and resilience. The intervention includes policies and programs to facilitate education, inclusion of vulnerable populations, and equal access to justice, quality healthcare and security. In this manner the programs and policies drive towards the goals as they directly address the causal agents.

The UN describes the Sendai Framework as somewhat of a departure from its previous work in that the focus has moved from managing disasters to reducing disaster risk and the framework sets forth definitive priorities and target goals [10]. One of the disasters that helped push this agenda was the Indian Ocean tsunami on Boxing Day, 2004 [14]. The tsunami was a result of an underwater earthquake at the Sumatra-Andaman fault that caused the ocean floor to rise by close to 40 ft. The force of this caused a change in the earth's rotation and triggered tsunami waves that reached heights of 100 ft and traveled at 500 miles an hour. The waves hit the coastlines of Thailand, India, and Sri Lanka first and then reached as far as South Africa, affecting a total of 10 countries. The loss of life was enormous, with estimates of almost 230,000 people killed and half as many injured. The tsunami destroyed coastline communities and displaced upwards of two million people. While the waves hit the coastlines of many Asian countries, the impact was felt around the world as many of the beaches were full of vacationers from other countries. In addition to the human impact, the economic impact was estimated to be close to $10 billion US dollars and recovery took years. The magnitude and widespread nature of both human and economic damage related to this tsunami likely served as a bellweather for change in the way nations and people approach disasters and influenced how the Sendai Framework strategic plan was structured and focused.

The Sendai Framework

The Sendai Framework follows the Hyogo Framework which was created to structure the work of the UNDRR relative to disaster management and risk reduction from 2005 to 2015 [9]. The Sendai Framework builds on the previous instruments by incorporating the concepts of resilience, prevention, and disaster management [10]. The product of a number of years of preparation through stakeholder interviews and negotiations with member nations reset the focus of the work and aligned it more with risk reduction, community resilience (including health structure), and the role of nations and local communities in engaging entire populations, including people with disabilities, to play an active part in the disaster management cycle, especially risk reduction. The Sendai framework will guide the next 15 years (2015 to 2030) of the UNDRR and member nations' disaster related work. The focus of inclusion, whole community and resilience parallels that of many nations across the globe. including Australia, the United States, and the European Union [15–17]. The Sendai framework sets out a focus and set of goals that incorporate the input and needs of people with disabilities and supports the UN's focus on disability inclusion [18].

The 2004 Indian Ocean tsunami led to significant loss of life, injury and economic damage including loss of homes and businesses. With the backdrop of the continued and increased impact of disasters, like this tsunami, the countries at the World Conference where the Sendai Framework was adopted expressed an urgency to move this work forward [9]. Evidence showed that disaster risk and exposure increased faster than resilience could be built or vulnerabilities diminished for many people supported this urgency. Thus, the Sendai Framework endeavors to slow and reverse this trajectory to begin to build the resilience needed to avoid the significant disaster impacts.

As the successor instrument to the Hyogo Framework for Action, the Sendai Framework benefited from the lessons learned in the decade preceding it [9,10]. The Hyogo Framework demonstrated that countries and regions made progress towards achieving disaster risk reduction at multiple levels. As well, it showed instances where deaths were reduced for some disasters and there was evidence of the cost-effectiveness of disaster risk reduction in preventing future economic losses.

The Hyogo Framework also contributed to the involvement of stakeholders and the development of partnerships which raised awareness of the issues and gained commitment from political institutions, both of which helped focus the activities. However, despite the progress, disasters still endanger sustainable development and result in significant losses of life and livelihood. In evaluating this situation, the Hyogo Framework decade identified some gaps in the framework's formulation. In particular, two areas were mentioned. One is the need to consider how to adequately implement disaster plans and the other is the need to support building resilience at all levels. With these successes and gaps, the Hyogo Framework set the stage for the Sendai Framework to push disaster risk reduction and nurture resilience building in individuals, communities, countries and globally through the impacts on and strengths of economies, social and cultural capitals, optimizing health and eco-system outcomes.

In moving forward from the previous decade and framework, the Sendai Framework takes a more expansive approach to the problem and aims to craft disaster risk reduction activities that address multiple hazards and multiple different populations [18]. This concept is referred to in the Sendai Framework as "multi-hazard" and "multisectoral," and is reflected in similar approaches with different names in other countries. Sometimes referred to as an "all hazard" approach to disaster management, this approach incorporates needs for multiple different kinds of disasters including natural disasters and those that are anthropogenic (human caused). This concept is reflected in the WHO's approach to health emergencies which considers that while emergencies may have different causes there is much that is similar in the emergency management approach and activities to address them [19]. It is also the approach taken by both the US Federal Emergency Management Agency (FEMA) and the Canadian Government's Public Safety entity [20,21]. Using this model streamlines planning and responding to disasters as it uses common themes and practices rather than a different process for every different kind of disaster. The latter approach generally proves to be inefficient as well as often ineffective.

Another concept guiding the actions of UNDRR that differentiates the Sendai Framework from the Hyogo Framework relates to engaging and involving people from all walks of life in the process [9]. The Sendai Framework uses the term "multisectoral" meaning that stakeholders include people from multiple sectors or areas of society with different interests, knowledge and needs [10]. The purpose of bringing together the private and public sectors along with academia and community based organiza-tions is to create opportunities to build networks, human and social capital to help foster resilience. In addition, the framework advocates for the inclusion of not only people in power, but also women, older people, people with disabilities, people living in poverty, and minority populations such as in-digenous peoples. The framework states that all of these stakeholders should be included in planning and policy implementation, response, recovery and mitigation. This concept is also reflected in the emergency management approach of many countries and country partnerships including the European Union, Australia and the United States, and is a focus of the UN and the WHO in general [3,16,22–24]. This concept, sometimes referred to as the "whole community approach" or "inclusive" emergency planning, has the same intent and meaning regardless of the term used.

The new direction of the Sendai Framework, with its clear focus on supporting and improving the health and economic well-being of people. complements the UN vision and mission to promote human rights and freedoms and combat poverty [2,9,10]. The attention to potentially vulnerable populations also shifts activities to those that build greater resilience. The method to reach the goals of preventing new and reducing existing disaster risks while promoting sustainable development uses the "capitals of community resilience" as defined by Emery and Flora [25]. These capitals represent the assets and resources of the community that contribute to its well-being. As well, many are areas that can be en-hanced and strengthened through inclusive community engagement. The Sendai Framework supports the employment of actions from various areas, including economic, political and legal, social and cul-tural, health, and environmental arenas with natural and built elements to increase resilience through the emergency management process. It notes that mitigating hazard risks and strengthening prepared-ness and capabilities for response and recovery through the planning process will expedite the building of resilience. In order to achieve this aim, the framework proposes seven global targets to reach by 2030 as measured by the difference between rates and indicators between the two decades 2005 to 2015 and 2020 to 2030 [9]. These target goals appear in Fig. 2.

1. Reduce the global rate of mortality related to disaster on average per 100,000 people.
2. Decrease the average number of people per 100,000 affected by disasters globally.
3. Reduce economic losses from disaster related to the global gross domestic product (GDP)
4. Foster resilience to significantly decrease disruption to basic services including healthcare facilities and schools and damage to critical infrastructure caused by disasters
5. Substantially increase support for developing nations by blending adequate and sustainable international resources with those of the nation at all levels
6. Significantly improve the availability and accessibility of disaster risk information, assessments, and early warning systems for multiple kinds of disasters
7. Substantively raise the number of countries with strategies for disaster risk reduction at the national and local levels*

*This target goal to be reached by 2020

FIG. 2

Seven global targets of the Sendai Framework.

Excerpted from the Sendai Framework. Sendai framework for disaster risk reduction; 2015. Available from: https://www.unisdr.org/ files/43291_sendaiframeworkfordrren.pdf.

These goals, although they don't have specific, numeric targets, address many of the issues that have plagued disaster management over time as well as recently. Despite the successes of previous instruments for achieving these goals, gaps were identified in the efficacy of the framework. The Sendai Framework benefits from the ability to learn from previous experiences, capitalizes on the advancement of the field, and incorporates those newer elements and focus into the framework.

United Nations disability and development

Preparation for the Sendai Framework occurred in the few years prior to the 2015 World Conference on Disaster Reduction, during a period of UN focus on disability issues and #Envision2030 initiative to transform the world into an inclusive place for people with disabilities [26,27]. Going by the moniker @UNEnable, the campaign outlines the 17 UN Sustainable Development Goals (SDG) developed for the world at large and advocates for disability inclusion in achieving the goals. The goals appear in Fig. 3. It is noted that there are a significant number of times that the SDG specifically refer to people with disabilities and consciously and deliberately include them in that goal [26]. There are seven places in five of the goals where people with disabilities are explicitly mentioned and those areas are highlighted in Fig. 3 [28]. As well, five goals, including education, economic growth and employment, reducing inequality, and accessible communities and homes, constitute important areas of attention for this campaign.

The 15 year lifespan of the SDG parallels that of the Sendai Framework, 2015 to 2030, and both have a concerted focus on disability inclusion [10,26]. The UN is working towards implementation of these SDG for all people, including those with disabilities, to mainstream disability into SDG implementation and raise disability awareness through #Envision2030. In addition, this campaign has targeted other key areas for focus, including the impact of the COVID-19 pandemic on disabled people, mainstreaming disability in development, and the monitoring and evaluation of disability-inclusive

1. No poverty	10. **Reduced inequality**
2. Zero hunger	11. **Sustainable cities and communities**
3. Good health and wellbeing	12. Responsible consumption and production
4. **Quality education**	13. Climate action
5. Gender equality	14. Life below water
6. Clean water and sanitation	15. Life on land
7. Affordable and clean energy	16. Peace and justice strong institutions
8. **Decent work and economic growth**	17. **Partnerships to achieve the goal**
9. Industry, innovation and infrastructure	

FIG. 3

17 UN sustainable development goals for 2030 [26,28]. (Bolded goals explicitly contain language including people with disabilities).

development in addition to operationalizing the 2030 SDG and working towards disability-inclusive, sustainable urban development [28,29]. In addition to these key areas of focus, the UN Department of Economic and Social Affairs disability group provides an array of toolkits and information on strategies and practical guidance about creating accessible environments at home and in the community, buying an accessible vehicle, accessible transportation, and other important areas.

In its technical note related to operationalizing the 2030 goals, the Secretariat stresses the importance of monitoring and evaluation of the SDG and disability-inclusive practices which come with opportunities and challenges [29]. This document harkens back to the 2000 Millennium Development Goals (MDGs) and the state of disability specific data and statistics at that time. At the time, there was no consistent way to capture this data across the world, making it difficult to understand the data or the disability populations that it represented. As well, the MDGs did not reflect disability specific issues or address their integration into working on the goals. In the meantime, much effort has gone into attempting to gather information related to people with disabilities in order to be able to understand the needs of this population.

Disability statistics and data

A number of groups, including UNICEF, the Washington Group on Disability Statistics (Washington Group), and the WHO have developed short surveys for countries to use to gather consistent data on people with disabilities, for use related to economic development, disaster risk reduction and building resilience, among others [29]. In particular, UNICEF worked with the Washington Group to create a set of child disability questions addressing function and disability in children 2 to 17 years old. The WHO developed its Model Disability Survey to capture information about people with varied levels of disability and people without disabilities in a way that allows for comparisons between these populations to identify gaps and needs [30]. Another tool developed by the Washington Group uses a small number of questions as an add-on to national censuses and surveys to provide a set of questions about the functional areas that are most likely to result in impairment and activity limitations in everyday life [31]. These six functional areas are vision, hearing, mobility, cognitive functioning, self-care and communication. Each question has answer choices reflecting functional ability, including whether the survey respondent performs the function with no difficulty, some difficulty, a lot of difficulty, or is unable to perform it. Two additional answer choices, "refused" and "don't know," don't provide a functional level. This tool can help stratify the disability population of a country by level of impairment as well as functional area of impairment. It also provides valid comparisons of populations across a country and across various countries regardless of culture or economic resources. It is available in multiple languages and estimated to take only about a minute of time to ask.

Data gathering is just a start as data without analysis and interpretation is not very informative. For data about people with disabilities to be used by policy or law makers, it must be understood and put into the context of the general population. For example, in its census of 2011, Mauritius captured data about employment to compare employment rates for people with and without disabilities [31]. Looking at the unemployment rate alone would have shown that only 3% of people with disabilities were unemployed compared to 5% of those without. However, this doesn't portray the full picture because it appears that only 26% of people with disabilities in Mauritius are employed, while 62% of those without are. What those statistics didn't appreciate was that 52% of the disabled population were not looking for work and therefore were not counted as unemployed even though they were not working. Many people with disabilities would like to work, but may not be able to due to barriers, such as lack of transportation and accessible workplaces. Some barriers to employment can be overcome. If the data is not analyzed in a way to take into account the potential differences between the general population and those with disabilities, then opportunities to overcome barriers to activities such as employment will be lost. Evaluating this data in a way that identified that 52% of people with disabilities were not looking for work allowed Mauritius to better understand the data and to begin to identify and address the barriers leading to disparities in a more effective manner. Making employment available to people with disabilities who want to work conveys benefits of improved economic status, sustainable development, mental health and well-being, all leading to greater levels of resilience and reduced disaster risk.

The UN focus on equality and sustainable economies as well as accessible communities and incorporates people with disabilities to help drive achieving these goals. Earlier documents, such as a focus report from the European Coalition of Community Living, outlines the principles of Article 19 of the UN Convention on the Rights of Persons with Disabilities (CRPD), stating that all people, including those with disabilities, have the right to live in and be active, valued and equal members of their community [32]. Part of participation in the community includes being a part of community planning, policy making, and disaster risk reduction. Studies show that the inclusion of people with disabilities in all levels of disaster management results in decreased vulnerability to the impacts of disasters and improved resilience [3]. Furthermore, people with disabilities bring a unique array of ideas, experiences and viewpoints to the table that have often not been considered, but may provide the optimal solution to the problem. Sometimes these solutions, like curb cuts in sidewalks, benefit not only people using wheelchairs and canes, but also those using strollers, scooters and bikes. Similar benefits for the whole community come from accessible schools, public transportation, and green and public spaces contribute to the health, wellness and inclusion of the whole community.

The World Health Organization

An agency of the UN, the World Health Organization (WHO) functions as its public health arm with the mission to promote the highest level of health for all people of the world [33]. Like other UN agencies, it serves to collaborate with the member nations and create partnerships and connections. The WHO's constitution stresses equity and inclusion as guiding principles for its work with a focus on using a scientific evidence base to develop policies. The WHO has the responsibility of addressing the UN's health-related SDGs as well as guiding the health aspects of disaster management and addressing the post disaster risks for communicable diseases and reconstruction and rehabilitation for the population.

The language used by various entities for the parts of the emergency management cycle differs somewhat. The WHO model, called the "disaster development continuum," uses a five part cycle with "preparedness," "response," "reconstruction," "rehabilitation," and "prevention", as seen in Fig. 4 [34]. According to the UN, rehabilitation refers to the process of empowering and supporting people with disabilities to reach their best possible level of functioning and independence [22]. Interventions include a variety of services and devices to restore or replace function, allowing the person to meet their needs. For example, someone that had a spinal cord injury may no longer be able to walk, but could be independently mobile using a wheelchair. Reconstruction constitutes the rebuilding of the physical, economic, and social infrastructure to meet the needs of the population. This includes assessing the need for disability services as well as reconnecting people with disabilities to their families and social networks. The WHO advocates for using the concept of "Design for All" when rebuilding community infrastructure, including public and green spaces, to make them accessible to all [34].

The language for the steps in the emergency management programs in Australia and some agencies in the United States differs somewhat with the use of a four step and five step model respectively [35,36]. Fig. 5 shows the five step US model with prevention, preparedness, response, recovery and mitigation. The Australian model with the four step cycle uses the same steps but without a separate mitigation step. Each of these two models and that of the WHO employ the terms "response," "preparedness" and "prevention." However, the differences lie in the other terms. Where the WHO

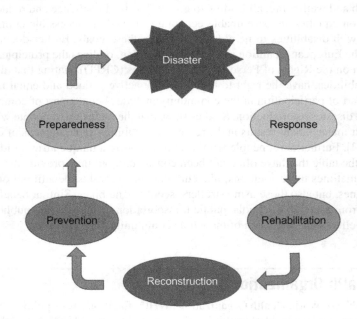

FIG. 4

The disaster-development continuum.

Adapted from WHO training Disasters and emergencies: definitions, training. World Health Organization; 2002 March.

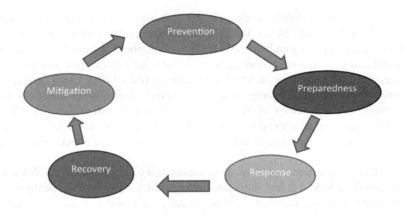

FIG. 5

The FEMA emergency management cycle [36].

model uses "reconstruction" and "rehabilitation," the other models use "recovery" or "recovery and mitigation" to refer to similar actions and functions.

WHO guidance for disaster planning and management mirrors the overall goals and mission of the UN in its approach to inclusive resilient communities. Access to medical care for acute and chronic conditions constitutes one of the elements that influence health and wellbeing. Other factors include genetics, behavior, environmental and physical influences and social factors [37]. These social determinants of Health (SDOH) can significantly influence the health of an individual or a community. Understanding the magnitude of impact of each of these factors on health and the intersections between them helps identify where to invest resources to make a difference.

Social determinants of health

The US Centers for Disease Control and Prevention pulled together the evidence for the relative impact of the various health determinants and notes that the effects of behavior and the social and physical environment have more influence on health outcomes than medical services [38]. Estimates of the relative magnitude of effect on health outcomes for each SDOH appear in Fig. 6. Individual behavior and the social and physical environment are often considered the non-medical determinants of health and account for more than half of the influence on health. As well, there is a predominance of intersection between these factors that further exacerbate health disparities between populations [39]. For example, tobacco related diseases including Chronic Obstructive Pulmonary Disease (COPD) and lung cancer

1. Medical care and services 10 to 20%
2. Genetics 30%
3. Behavior 40 to 50%
4. Social and physical environment 20%

FIG. 6

Impact on health outcomes of social determinants of health [38].

which kill more than 8 million people worldwide; 80% of tobacco users come from low and middle income nations [40]. In the United States, tobacco use is more common in lower income and minority populations and prevalent in the population of people with serious mental illness [39]. It is also very addictive and a difficult habit to break, making smoking or other tobacco use cessation difficult. Long term tobacco use, especially smoking, leads to lung disease like COPD. COPD creates disability by limiting functional abilities related to lung capacity and causes disability for many due to shortness of breath during activities of daily living (ADLs), limits physical activities and causes subsequent physical fitness deconditioning and worsened muscle strength, all of which leads to functional limitations [41]. Thus, behavior, income and educational level impact tobacco use which in turn leads to poorer health for tobacco users.

Health behavior, such as tobacco use, makes up almost half of the impact on health status. Other behaviors affecting health include diet and nutrition, both of which contribute to obesity and malnutrition depending on the population, alcohol consumption, sexually transmitted disease, and preventive screening for chronic diseases and risk factors [38,39,41]. These health behaviors are similar between developed and developing countries, although the outcomes may be different and influenced by the social, political and economic factors of the country. Diet and nutrition exemplifies this point in that the developed world battles problems with overweight and obesity, while the developing world still struggles with malnutrition and difficulty feeding the population. Developing countries lack resources for screening for chronic treatable conditions and risk factors such as high blood pressure that untreated lead to heart disease and diabetes with complications of heart disease, peripheral circulation difficulties, visual dysfunction and kidney disease [42]. Primary care provides screening for chronic conditions and risk factors and early treatment and prevention of the condition or complications [43]. Prevention and early treatment result in better health outcomes and more cost effective care with less need for inpatient hospital use. However, barriers to this approach exist in both developing countries and rural areas in developed countries. Both areas often have limited access to primary care and preventative health screenings because of healthcare professional shortages and long travel distances to reach healthcare services [42,44].

Health literacy, linked to more positive health outcomes, is lower in many people with low to middle incomes, living in poverty, and living in rural and underserved areas [45]. These factors have an even greater impact on people with disabilities. The literature shows that people with disabilities are less likely to receive quality healthcare, including preventive care, than those without disabilities [46–49]. This is true for multiple disability populations, including those with intellectual and developmental disabilities, physical disabilities, and older adults with chronic conditions. A study of people with physical disabilities show that they are less likely to have a high quality usual source of medical care and that this disparity is more prevalent in minorities than whites [48]. In their study, Mahmoudi and Meade identified that medical care providers for older people with physical disabilities and those with low incomes referred them for smoking cessation services less frequently than their other patients. They proposed that policies with financial incentives to support a usual source of medical care and appropriate access to behavioral interventions would mitigate some of the disparities and improve health. Lack of access to quality medical care and the inequities that it creates for people with disabilities contribute to poorer health and greater vulnerability of this population.

In 1977, the World Health Assembly created a resolution for a "Health for All" approach to address social determinants of health and health inequities [50]. The WHO European Healthy Cities Network (WHO-EHCN) uses this standard to drive its work in developing policies and programs to decrease inequalities. Researchers interested in identifying which of the social determinants most affects health have looked for the underlying causes of social determinants [51]. After examining factors that contribute to health and what is known about biological pathways, Braveman and Gottlieb determined that the primary causes of a broad array of health outcomes are linked to socioeconomic factors including education, income, and wealth. In recognition of this finding and from their decades of work to attempt to achieve health equity, the WHO-EHCN shifted their focus from policies and programs directly affecting health for vulnerable populations to looking upstream at interventions that address the underlying causes. The more recent policy challenge seeks to address socioeconomic determinants of health such as poverty, unemployment, education and housing [50]. This work circles back to the basic mission of the UN related to improving the lives of people living in poverty by addressing hunger, illiteracy, and disease [1]. By addressing poverty, the UN and its agency the WHO will also improve health outcomes for people living in poverty without access to quality medical care including those with disabilities. Reducing vulnerabilities for these populations and improving health will decrease their disaster risks and improve their resilience.

Resilience represents a tactic to reduce the impact of disasters on individuals and communities and has become an important focus for disaster management. While health status contributes to an individual's resilience, there are other factors that also promote resilience. Social networks embody one of the most important factors in resilience [25]. Social participation, including belonging to social groups, involvement in religious activities and volunteering, are critical to building and maintaining social networks. Studies in Ghana identify that social networks can also be important in building health literacy even in a population of people that is socioeconomically disadvantaged [52]. People with high levels of social participation build their health knowledge and literacy through interactions with others, which in turn leads to better health and well-being. The author of the Ghana study proposes that social participation be channeled to improve the health and health literacy of a population. This public health approach to improving health outcomes would also contribute to building resilience to disasters and is likely to be effective for people with disabilities as well those in the general population.

Mental health and disasters

The WHO recognizes the importance of addressing mental health as well as physical health [53]. It estimates that about 22% of people living in areas that have experienced conflict will have a mental illness such as depression, anxiety, post-traumatic stress disorder (PTSD), bipolar disorder or schizophrenia. Many people experience some symptoms of depression, anxiety or PTSD after a disaster, however, for most people those resolve over time. For people with serious mental illness such as bipolar disorder, major depression or schizophrenia that preceded the disaster, recovery may be more difficult. As well, they often contend with poorer physical health in addition to their mental illness [54]. People with disabilities and chronic medical conditions have a higher prevalence of mental illness than the general population, making access to mental health support and treatment important. Effective approaches to mental health during and after a disaster require access to a range of mental health services to support people during the disaster and hopefully avoid the development of longer lasting complications [53]. These services range from psychological first aid for acute needs during and immediately after the disaster to care from a mental health professional. The list of mental health services and supports identified by the WHO appear in Fig. 7. As for other resources and services, disasters can provide the opportunity to build a stronger, more robust mental health system post-disaster. In the aftermath of the Indonesian tsunami in 2004, Sri Lanka made mental health services a priority and identified a shortage

1. Community self-help and social support-lay person support and problem solving
2. Psychological first aid-first line support from trained lay people
3. Basic clinical mental health care by general health personnel
4. Psychological interventions such as cognitive behavioral therapy or group therapy
5. Psychiatric hospitalization with psychiatry and other specialized staff

FIG. 7

WHO-endorsed interagency mental health and psychosocial support guidelines.

Adapted from NIMH. Chronic illness and mental health: recognizing and treating depression. www.nimh.nih.gov; 2021. Available from: https://www.nimh.nih.gov/health/publications/chronic-illness-mental-health.

of mental health resources. The country developed an initiative to increase the availability of mental health services and built resources to provide them. While prior to the initiative only 10 out of its 27 districts had mental health resources, after its completion those services were available in 20 of the 27 districts. These lessons illustrate that by prioritizing resources needed in a disaster, it is possible to develop them and to build back a better community. Communities and nations create better solutions by learning from these examples and proactively incorporating people with mental health needs and mental health service providers in inclusive emergency planning.

International disability disaster programs

Many nations including developing ones have begun to make the inclusion of people with disabilities in disaster planning a priority. The UN, WHO and other groups provide direction and resources for these activities, but in order to meet local needs the bulk of the task falls to the nations, their people and local communities. In some cases, developed and developing countries have partnered to build awareness of and plan for how to incorporate people with disabilities into disaster risk reduction [55]. The UNDRR has joined with foundations to commit to making the principles and details of the Sendai Framework available to people with disabilities. In addition, developing countries have paired with universities in developed countries to learn more about disaster risk reduction and in particular what people with disabilities need to know as well as how to meaningfully engage in disaster risk reduction efforts. For example, the University of Sydney's Centre for Disability Research and Policy has worked with Cambodia, the Philippines, and Thailand to create a toolkit to teach people with disabilities about disaster risk awareness and preparedness using people with disabilities as the trainers. The Centre also partners with its own country, Australia, in addition to its Asia-Pacific work, with a goal towards the optimization of health and wellness in people with disabilities and their engagement and participant in the social and economic activities within their communities [56]. This work has developed the Person-Centered Emergency Management framework to guide the implementation of disability inclusion in all aspects of disaster management. Other countries have embarked on projects to include people with disabilities in disaster risk reduction and their work reflects both grass-roots and governmental policy strategies. The activities of two countries illustrate these different kinds of activities and approaches that lead to whole community involvement and input of people with disabilities into disaster risk reduction activities.

Bangladesh employs an inclusive planning process through its Community Based Disability Inclusive Disaster Risk Reduction project in the Gaibandha district [55]. In this project, Bangladesh introduced a disability component into all levels of disaster risk reduction and planning. The Brahmaputra River which runs through Gaibandha floods on an annual basis, creating a regular pattern of destruction. In order to reduce flood risk, the district brought together the community to identify ways to reduce the impact of flooding. People with disabilities were included in this group and asked to provide input on their experiences and needs. They also received risk assessment training as part of the project. Building from project activities, the villagers drew a map of the community that showed areas of flooding and the location of critical infrastructure and evacuation routes. The map also contained location information for elderly villagers' homes, the homes of people with disabilities, livestock, and other village resources. This map allowed them to plan where to place evacuation shelters so they would be out of the flood plain, how to store and protect food and other resources, where to put evacuation routes, and community roles responsible for moving people and livestock away from flooding areas. The input of the group, especially that of those with disabilities, provided a valuable tool that helped the villagers formulate their evacuation plans to protect both their people and their resources.

Ecuador also approached concerns about meeting the needs of disaster needs of disabled people from a higher level. They formed a governmental entity, the Technical Secretariat for the Inclusive Management on Disabilities (SETEDIS), to address the needs of this population. The SETEDIS tracked the whereabouts of known people with disabilities after the 2016 earthquake in northern Ecuador [57]. Events like earthquakes often affect a region that encompasses multiple countries and Ecuador partnered with nearby countries to discuss regional integration of disaster management. In response to the principles set forth in the Sendai Framework, this partnership discussed the role of people with disabilities in disaster risk reduction planning [55,57]. As part of the country's actions in disability-inclusive disaster management, Ecuador developed a process led by SETEDIS to identify and geo-locate people with disabilities in order to determine where to put resources to meet their needs. This included ensuring that shelters and other resources are accessible to people with disabilities so they can stay in their community.

Conclusion

As part of its mission to improve the lives of people living in poverty, the UN tasked two of its agencies, the UNDRR and the WHO with addressing the needs of people with disabilities in the face of disasters. The disaster reduction frameworks of the World Conferences have progressively become more inclusive in concept and focused on prioritizing engagement and participation of people with disabilities in the process to better meet their needs. Both the UNDRR and the WHO participate in supporting inclusive disaster risk reduction through various activities. In addition, the WHO, as a public health entity, promotes health by addressing social determinants of health and their underlying causes that lead to disproportionally poorer health in those with disabilities. This work tackles the lack of access to medical care and health behaviors as well as socioeconomic and physical environment issues. Their work also recognizes the importance of the availability of both physical and mental health resources after a disaster and to support recovery from and complications of the physical and psychological impacts. All of this work contributes to building resilience and capacity in the populations of people with disabilities and in their communities. Furthermore, the emphasis on inclusion of people with disabilities in their communities and in disaster management and risk reduction processes has turned the spotlight on this priority. The resulting actions of countries and partnerships have produced valuable and innovative work and partnerships resulting in stronger roles for people with disabilities in their community. This is a beginning as there is clearly more progress needed in the meaningful inclusion of people with disabilities in community disaster management and risk reduction in both developed and developing countries.

Questions for thought

1. Choose a country in the world and a type of disaster event and outline how to incorporate people with disabilities into disaster management in their community or at a regional or national level.
2. Select one of the 17 UN Sustainable Development Goals for 2030 and discuss how it might impact people with disabilities with regard to disaster response and recovery.
3. Identify how a physical or mental health system could be employed during a disaster to better meet the needs of people with disabilities.

References

[1] United Nations Seventieth Anniversary. (2015). *History of the UN.* United Nations. [cited 2022 Jan 7]. Available from: https://www.un.org/un70/en/content/history/index.html.
[2] General Assembly of the United Nations. (2022). *Vision statement.* United Nations. [cited 2022 Jan 7], Available from: https://www.un.org/pga/74/documents/vision-statement/.
[3] *Disability-inclusive disaster risk reduction and emergency situations.* (2015). United Nations Enable. www.un.org. Available from: https://www.un.org/development/desa/disabilities/issues/disability-inclusive-disaster-risk-reduction-and-emergency-situations.html.
[4] *History.* www.undrr.org. Available from: https://www.undrr.org/about-undrr/history.
[5] *United Nations General Assembly session 44 resolution 236. A/RES/44/236.* (22 December 1989).
[6] Notice to readers international decade for natural disaster reduction. (1994). *MMWR, 43*(17), 321–322.
[7] *Australian multilateral assessment.* (2012 March). Australian Government. [cited 2022 Jan 7]. Available from: https://www.aph.gov.au/.
[8] *Words into action: practical guidance on implementing a people-centered approach to DRR.* (2022). United Nations. [cited 2022 Jan 7]. Available from: https://www.undrr.org/.
[9] *Hyogo framework for action 2005-2015: building the resilience of nations and communities to disasters.* (2005). Available from: https://www.unisdr.org/2005/wcdr/intergover/official-doc/L-docs/Hyogo-framework-for-action-english.pdf.
[10] *Sendai framework for disaster risk reduction.* (2015). Available from: https://www.unisdr.org/files/43291_sendaiframeworkfordrren.pdf.
[11] UNISDR. (2016). *Strategic framework.* Available from: https://www.unisdr.org/files/51557_unisdrstrategicframework20162021pri.pdf.
[12] *Theory of change UNDAF Campanion guidance 2 ! UNDAF companion guidance: theory of change.* Available from: https://unsdg.un.org/sites/default/files/UNDG-UNDAF-Companion-Pieces-7-Theory-of-Change.pdf.
[13] *Theory of change concept note I. What is a theory of change?.* (2016). [cited 2022 Jan 7]. Available from: https://unsdg.un.org/sites/default/files/16.-2016-10-18-Guidance-on-ToC-PSG-LAC.pdf.
[14] Reid, K. (2019). *2004 Indian ocean earthquake and tsunami: facts, FAQs, and how to help.* World Vision. Available from: https://www.worldvision.org/disaster-relief-news-stories/2004-indian-ocean-earthquake-tsunami-facts.
[15] *Australian disaster preparedness framework: a guideline to develop the capabilities required to manage severe to catastrophic disasters.* (2018). Available from: https://www.homeaffairs.gov.au/emergency/files/australian-disaster-preparedness-framework.pdf.
[16] FEMA. (2011). *A whole community approach to emergency management: principles, themes, and pathways for action.* Available from: https://www.fema.gov/sites/default/files/2020-07/whole_community_dec2011__2.pdf.
[17] Hans. (2020). *European disaster risk management.* European Civil Protection and Humanitarian Aid Operations - European Commission. Available from: https://ec.europa.eu/echo/what/civil-protection/european-disaster-risk-management_en.

[18] United Nations. (2006). *Convention on the rights of persons with disabilities (CRPD)*. United Nations. Available from: https://www.un.org/development/desa/disabilities/convention-on-the-rights-of-persons-with-disabilities.html.

[19] *Policy*. www.euro.who.int. (2021). Available from: https://www.euro.who.int/en/health-topics/health-emergencies/from-disaster-preparedness-and-response/policy.

[20] *National preparedness guidelines*. (2007). Available from: https://www.fema.gov/pdf/emergency/nrf/National_Preparedness_Guidelines.pdf.

[21] Public Safety Canada. (2018). *All-hazards risk assessment*. www.publicsafety.gc.ca. [cited 2022 Jan 7]. Available from: https://www.publicsafety.gc.ca/cnt/mrgnc-mngmnt/mrgnc-prprdnss/ll-hzrds-rsk-ssssmnt-en.aspx.

[22] World Health Organization. (2021). *Disasters, disability and rehabilitation*. Available from: https://www.who.int/violence_injury_prevention/other_injury/disaster_disability2.pdf.

[23] *2.4.8. Strengthening IE with attention to people with disabilities*. (2018). Capacity4dev. europa.eu. [cited 2022 Jan 7]. Available from: https://europa.eu/capacity4dev/rnsf-mit/wiki/248-strengthening-ie-attention-people-disabilities.

[24] Villeneuve, M. (2021). Building a roadmap for inclusive disaster risk reduction in Australian communities. *Prog Disaster Sci, 10*, 100166.

[25] Emery, M., & Flora, C. (2006). Spiraling-up: mapping community transformation with community capitals framework. *Community Dev, 37*(1), 19–35.

[26] United Nations. (2015). *Envision2030: 17 goals to transform the world for persons with disabilities*. United Nations Enable. Un.org. Available from: https://www.un.org/development/desa/disabilities/envision2030.html.

[27] *Transforming our world: the 2030 Agenda for Sustainable Development, 70th session*. (2015). undocs.org. Available from: https://undocs.org/A/RES/70/1.

[28] *Operationalizing the 2030 Agenda: Ways forward to improve monitoring and evaluation of disability inclusion. Technical notes of the secretariat*. (2015 December).

[29] *Issues*. (2022). United Nations Enable. www.un.org. Available from: https://www.un.org/development/desa/disabilities/issues.html#health.

[30] WHO. (2017). *Model Disability Survey (MDS)—survey manual*. www.who.int. [cited 2022 Jan 7]. Available from: https://www.who.int/publications/i/item/9789241512862.

[31] The Washington Group on Disability Statistics. (2022). *Question sets*. [cited 2022 Jan 7]. Available from: https://www.washingtongroup-disability.com/question-sets/.

[32] (2009 August). *Focus on Article 19 of the UN Convention on the Rights of Persons with Disabilities. European Coalition for Community Living Focus Report*. [cited 2022 Jan 7]. Available from: http://community-living.info/wp-content/uploads/2014/02/ECCL-Focus-Report-2009-final-WEB.pdf.

[33] WHO. (2021). *About WHO*. Who.int. Available from: https://www.who.int/about.

[34] World Health Organization. (2002 March). *Disasters and emergencies: definitions, training*.

[35] Australian Disaster Resilience Knowledge Hub. (2021). *Emergency planning handbook*. knowledge.aidr.org.au. [cited 2022 Jan 7] Available from: https://knowledge.aidr.org.au/resources/emergency-planning-handbook/.

[36] FEMA—Emergency Management Institute (EMI). (2018). *National Preparedness Directorate National Training and Education Division*. Fema.gov. Available from: https://training.fema.gov.

[37] *Closing the gap in a generation: health equity through action on the social determinants of health. Commission on Social Determinants of Health*. (2008). Geneva: World Health Organization.

[38] *CDC—Social Determinants of Health*. (2021). www.cdc.gov. Available from: http://www.cdc.gov/socialdeterminants/Definitions.html.

[39] *Institute of Medicine (US) Committee on health and behavior: research, practice, and policy, health and behavior: the interplay of biological, behavioral, and societal influences*. (2001). Washington, DC: National Academies Press (US).

[40] World Health Organization. (2021). *Tobacco*. Who.int. World Health Organization. Available from: https://www.who.int/news-room/fact-sheets/detail/tobacco.

[41] Regnault, A., Aguilaniu, G.-B., Dias-Barbosa, C., Arnould, M., et al. (2011). Disability related to COPD tool (DIRECT): towards an assessment of COPD-related disability in routine practice. *Int J Chron Obstruct Pulmon Dis*, (July), 387.

[42] Eshetu, E. B., & Woldesenbet, S. A. (2011). Are there particular social determinants of health for the world's poorest countries? *Afr Health Sci*, *11*(1), 108–115.

[43] Muzammil, S., & Muzammil, S. (2021). *Lack of primary health care services in developing countries during pandemic: an urgent reminder!*. www.clinmedjournals.org. [cited 2021 Mar 31] Available from: https://www.clinmedjournals.org/articles/jfmdp/journal-of-family-medicine-and-disease-prevention-jfmdp-7-138.php?jid=jfmdp.

[44] Rural Health Information Hub. (2019). *Healthcare access in rural communities introduction—Rural Health Information Hub*. Ruralhealthinfo.org. Available from: https://www.ruralhealthinfo.org/topics/healthcare-access.

[45] Marrocco, A., & Krouse, H. J. (2017). Obstacles to preventive care for individuals with disability. *J Am Assoc Nurse Pract*, *29*(5), 282–293.

[46] Anderson, L. L., Humphries, K., McDermott, S., Marks, B., Sisirak, J., & Larson, S. (2013). The state of the science of health and wellness for adults with intellectual and developmental disabilities. *Intellect Dev Disabil*, *51*(5), 385–398. https://doi.org/10.1352/1934-9556-51.5.385. Erratum in: Intellect Dev Disabil 2013 December;51(6). doi:10.1352/0047-6765-51.6.fmii.

[47] Mahmoudi, E., & Meade, M. A. (2015). Disparities in access to health care among adults with physical disabilities: analysis of a representative national sample for a ten-year period. *Disabil Health J*, *8*(2), 182–190. https://doi.org/10.1016/j.dhjo.2014.08.007.

[48] Miller, N. A., Kirk, A., Kaiser, M. J., & Glos, L. (2014). Disparities in access to health care among middle-aged and older adults with disabilities. *J Aging Soc Policy*, *26*(4), 324–346. https://doi.org/10.1080/08959420.2014.939851.

[49] Ritsatakis, A. (2012). Equity and the social determinants of health in European cities. *J Urban Health*, *90*(S1), 92–104. Available from: https://www.ncbi.nlm.nih.gov/pmc/articles/PMC3764273/. https://doi.org/10.1007/s11524-012-9762-y.

[50] Braveman, P., & Gottlieb, L. (2014). The social determinants of health: it's time to consider the causes of the causes. *Public Health Rep*, *129*(Suppl. 2), 19–31. https://doi.org/10.1177/00333549141291S206.

[51] Amoah, P. A. (2018). Social participation, health literacy, and health and well-being: a cross-sectional study in Ghana. *SSM - Popul Health*, *4*(April), 263–270.

[52] World Health Organization (WHO). (2019). *Mental health in emergencies*. Who.int. World Health Organization: WHO. Available from: https://www.who.int/news-room/fact-sheets/detail/mental-health-in-emergencies.

[53] NIMH. (2021). *Chronic illness and mental health: recognizing and treating depression*. www.nimh.nih.gov. Available from: https://www.nimh.nih.gov/health/publications/chronic-illness-mental-health.

[54] *Disability inclusion in disaster risk management*. (2018). GFDRR www.gfdrr.org. [cited 2022 Jan 7]. Available from: https://www.gfdrr.org/en/publication/disability-inclusion-disaster-risk-management-0.

[55] *Disability and the Japan and Ecuador earthquakes*. (2016). United Nations Enable. www.un.org. [cited 2022 Jan 7]. Available from: https://www.un.org/development/desa/disabilities/news/dspd/disability-the-japan-and-ecuador-earthquakes.html.

[56] Villeneuve, M., Abson, L., Pertiwi, P., & Moss, M. (2021). Applying a person-centred capability framework to inform targeted action on disability inclusive disaster risk reduction. *Int J Disaster Risk Reduct*, *52*(January), 101979. https://doi.org/10.1016/j.ijdrr.2020.101979.

[57] The University of Sydney. (2022). *Centre for disability research and policy*. [cited 2022 Jan 7]. Available from: https://www.sydney.edu.au/medicine-health/our-research/research-centres/centre-for-disability-research-and-policy.html.

The impacts of disasters on people with disabilities and chronic physical and mental health conditions

Susan Wolf-Fordham[a,b] **and Jill Morrow-Gorton**[c]

[a]*Consultant, Association of University Centers on Disabilities, Silver Spring, MD, United States,* [b]*Adjunct Faculty of Public Health, Massachusetts College of Pharmacy and Health Sciences, Boston, MA, United States,* [c]*University of Massachusetts Chan Medical School, Worcester, MA, United States*

Learning objectives

- Describe the potential impact of disasters, emergencies and pandemics like COVID-19 on the *health* of people with disabilities
- Describe the potential impact of disasters, emergencies, and pandemics like COVID-19 on the *daily lives* of people with disabilities
- Analyze the potential causes of those disaster impacts
- Describe the impact of emergencies, disasters, and pandemic on traditionally marginalized populations other than those with disabilities

Integrating Mental Health and Disability Into Public Health Disaster Preparedness and Response
https://doi.org/10.1016/B978-0-12-814009-3.00007-6

Introduction

It is well known that emergencies, disasters and pandemics disproportionately and negatively impact disabled people. These impacts can be viewed from several different viewpoints, such as health disparities, disruptions to everyday functioning, and barriers to equal access. Emergency planning and response falls into the realm of public health and as such constitutes a health issue. Evidence shows that disasters can cause disabled people to experience the worsening of existing medical conditions or the development of new conditions to a greater extent than the general population. This disproportionate health impact manifests as a health disparity. "Health and health care disparities refer to differences in health and health care between groups that stem from broader inequities" [1]. Disasters can also be considered in terms of the extent to which they cause disruptions to people's everyday lives. What moves an event from being an everyday occurrence, to an annoyance, to an emergency is the amount of disruption it causes. This disruption represents challenges to participating in usual everyday activities when an emergency renders services, resources, equipment, supports and the built environment inaccessible or unavailable. Such disruptions impact the ability to work, to manage health, and to participate in family and the community. These impacts can also be seen in terms of lack of access. In addition to the disruption of daily life, access to emergency services and facilities such as mass care disaster shelters (shelters for the general public), emergency warnings and alerts, evacuation and other services becomes more challenging to disabled people than to the general public. While disaster services are meant to serve the whole community, the planning needed to make these resources accessible to disabled people is impeded by a lack of emergency management and public health preparedness personnel: (1) awareness and knowledge of disability; (2) understanding how to implement the Americans with Disabilities Act of 1990 (ADA), a civil rights law, in emergencies; and (3) comfort level with stakeholder engagement. Examining the state of knowledge about disabled people within the emergency management system and understanding the historical impact that this has had in terms of health, disruption to daily life and access to resources during and after a disaster, lays a groundwork for finding ways to improve planning, response, mitigation, recovery and building resilience in this population and the whole community.

Health disparities and negative health impacts from disasters

Similar to people from other traditionally marginalized groups, children and adults with disabilities experience poorer health than the general population as well as health disparities. As evidence of poorer health, studies show that people with disabilities are more than four times more likely to rate their health as fair or poor than people without disabilities. In the United States (US) 40.3% of people with disabilities rated their health as fair or poor compared to 9.9% of people without disabilities [2]. These numbers are comparable to those reported by the United Nations (UN) for 43 different countries where 42% of people with disabilities rated their health as poor compared to only 6% of people without disabilities [3]. People with disabilities also experience health disparities which represent avoidable differences in population level health outcomes that can be related to social, economic, or environmental disadvantages. In the United States the term "health disparities" refers to avoidable and unjust causes, and "health differences" refers to both avoidable and unavoidable causes. In contrast, the World Health Organization (WHO) uses "health inequities" for avoidable and unjust causes and "health inequalities" to encompass both avoidable and unavoidable health differences [4]. For people with disabilities, some unavoidable health differences may exist related to the underlying cause of their disability. Nonetheless, the impact of avoidable causes is great, and the UN describes that people with disabilities largely have more healthcare needs related to both common and disability-unique conditions than the general population, making them more vulnerable to lack of access to healthcare services and the effect of low quality healthcare [3].

These vulnerabilities, caused by the combination of avoidable and unavoidable factors affecting health, magnify during a disaster. Those with disabilities and older Americans are up to four times more likely to die or experience serious injuries due to emergencies than people without these conditions [5]. During emergencies, disasters and pandemics, existing disabilities and medical conditions often worsen, exacerbated by the interruption in home management of health conditions, loss of access to health care, social services and other supports. Additionally, emergencies may lead to new disabilities or new medical conditions as a result of injury or complications related to an inadequately managed and treated chronic condition [6]. These changes negatively impact health, increasing the complexity of medical problems, care and management and may even lead to inappropriate institutionalization such as nursing facility admission [7,8].

Involuntary and inappropriate institutionalization has been reported in conjunction with a number of disasters. For example, in 2017, six people experiencing homelessness in Florida refused to evacuate during Hurricane Irma and were involuntarily committed under the Baker Act, a Florida law from the 1970s that permitted the involuntary commitment of people suspected of mental illness and homeless people who refused to voluntarily go to shelters [9]. The National Council on Disability (NCD) issued a report about institutionalization during disasters and emergencies that described a trend of placing disabled people who had lived in the community into institutional settings because of lack of accessible housing or challenges with disaster recovery [10]. The report concludes that the issue is further complicated by Medicare and Medicaid regulations which allow moving patients into nursing homes absent worsening healthcare needs requiring that level of care. The NCD report also notes that once a person has been institutionalized it is often challenging to return to the community despite existing rules and practices promoting community dwelling for people with disabilities.

Multiple federal laws and activities support community living for people with disabilities, including the ADA [11] which requires people with disabilities to be accommodated in the most integrated setting appropriate to their needs [10]. The 1999 US Supreme Court's decision in *Olmstead v. L.C.* (Olmstead), found the segregation of people with disabilities in institutions unjustified and a form of unlawful discrimination [10]. The Centers for Medicare & Medicaid Services (CMS) have made efforts to rebalance services for people with disabilities by expanding and enhancing home and community based services to allow people to move from institutional settings into the community [12,13].

COVID-19

Disasters such as the COVID-19 pandemic also disproportionately affect people with disabilities. In addition to facing the issue of virus spread, people with disabilities may be negatively impacted by a lack of equal access to COVID-related health care and local public health and emergency services. The negative impacts of COVID-19 on the health of disabled people are widely reported. A study of approximately 65 million people in the United States found that people with intellectual disabilities were about three times more likely to be diagnosed with COVID-19 than the general population, more likely to be admitted to the hospital, more likely to have hospital Intensive Care Unit (ICU) stays and more likely to die from COVID-19 [14]. In Washington state, a 13.5% death rate, almost triple the state's rate at the time, was reported for people with intellectual disabilities with COVID-19 living in state supported group homes. Similar numbers were reported in Oregon [15]. A study of insurance claims data found that people with developmental disabilities under age 70 were almost eight times more likely to die from COVID-19 than people without developmental disabilities [16].

Researchers have found that people with Down Syndrome also experience more complications with COVID-19 and they are 5 times more likely to need hospital care and 10 times more likely to die from COVID-19 than the general population [17]. Because people with Down syndrome have immune deficiencies, the Centers for Disease Control and Prevention (CDC) approved third vaccine doses for this group to boost their COVID-19 immunity during the pandemic. Studies show that this group experienced more breakthrough infections (occurring despite vaccination), than the general population [18]. This risk to people with Down syndrome exceeded that of other common risks factors for COVID-19 including nursing facility residence, cancer treatment with chemotherapy, and HIV/AIDS by more than three times. Similarly, the CDC considers other groups of people with chronic conditions at greater risk for complications of COVID-19 infections [17,19]. These groups include people with kidney, liver, heart or lung disease, dementia or other neurological conditions, diabetes, HIV/AIDS, immunodeficiencies from various conditions, mental health conditions including schizophrenia and depression, obesity and sickle cell disease. Nursing home residents, all of whom generally have chronic disabling physical or mental health conditions, have particular vulnerability to COVID-19. The COVID Tracking Project at *The Atlantic* reported that as of March 2021, deaths in long term care facilities comprised more than one third of US COVID-19 related deaths [20].

Disruptions to daily life

Many disabled people rely on a variety of supports and services, including assistive or other technology, medical equipment, service animals, and formal or informal care and support in their daily lives. Other people with disabilities depend on a stable and predictable daily routine to maintain their physical and mental health. Loss of access to daily routines, familiar environments, critical support services, and adaptive equipment happen commonly during disasters and separation from vital equipment and support networks often leads to loss of independence [8,21]. For example, people who need help with daily living activities like eating, bathing, grooming and dressing are dependent on the people who provide personal care to complete these tasks. If a severe storm closes roads and makes driving hazardous, so that their care providers are unable to reach them, a disabled person who receives these services would go without, potentially leading to health and hygiene problems and loss of dignity. Other services, including healthcare, mental and behavioral health appointments, and social or vocational services, work and day programs, may be canceled because of closures due to emergencies.

Similarly, disruptions to daily life and services impact people with intellectual disabilities and those with mental illness when the daily support that they typically receive is absent. The absence of these services may lead to increased behavioral issues that may be treated with psychotropic medication or require hospitalization rather than being addressed in their usual manner with daily support or services, or may lead to medication side effects such as weight gain and obesity [14].

In addition to reliance on support from other people, many people with disabilities use equipment that requires electric power to operate. Such equipment includes ventilators, power wheelchairs and scooters, oxygen, home monitoring equipment and refrigeration for storing medications. Power outages, both planned and due to emergencies, are problematic for these individuals. The loss of electric power can cause these devices to stop working, resulting in loss of mobility, loss of respiratory support, and the spoilage of refrigerated medications, all of which put the person's health at risk. People who use computerized augmentative and alternative communication (AAC) devices to communicate may lose their ability to communicate if they are unable to charge their device. Fig. 1 shares personal experiences from a report by Disability Rights Texas describing the health impacts on two people with chronic medical problems and disabilities due to lost power during a severe winter storm in 2021 [22].

Even planned power outages may be problematic. During wildfire season in October 2019, wildfires swept across California. The power companies used planned black outs in an attempt to prevent power lines blown by the wind from starting fires. While thousands of Californians prepared for the power outages, the outages impacted those with disabilities particularly hard [23]. The *Los Angeles Times* reported about a woman with a chronic genetic disease and impaired immune system whose health decompensated when she could not follow her usual routine. She noted that the 2 days notice before the power outages didn't always provide sufficient time to develop alternative plans for personal care or to ensure that all needed equipment was charged in preparation for an outage [23]. Advance planning for these and many other disruptions can prevent these negative health effects. For example, if the woman in the newspaper story above had had a longer notice period perhaps she could have better planned to limit her risk.

Lack of equal access

The COVID-19 pandemic has highlighted the health disparities and differences in access to healthcare for people with disabilities [24]. Not only were people with disabilities more greatly impacted by COVID-19 infections, but they also experienced reduced access to basic healthcare services during the pandemic. When healthcare visits morphed from in-person to telehealth, many people with disabilities

"My son takes a medication that has to be prescribed and handed to me. We had to ration my son's medication because I could not get the medication during this emergency."

"I receive home health services because I need assistance getting out of bed, toileting, eating. My agency called me two days before the storm to ask if I had help because they could not guarantee that my provider would make it in. Thankfully, I'm really good friends with my provider and she made it through to me. But not once did anyone call to see if they could help or assist me in any way. And if it had not been for her, I would have been stuck in bed without help for four days."

FIG. 1

Health Impacts from Winter Storm Uri.

Disability Rights Texas. The Forgotten Faces of Winter Storm URI the Impact on Texans with Disabilities When We Fail to Conduct Inclusive Disaster Planning and Preparedness. The Protection and Advocacy Agency for Texans with Disabilities; 2021 [cited 2021 Dec 21]. Available from: https://media.disabilityrightstx.org/wp-content/uploads/2021/04/06100917/apr-5-2021-DRTX-winter-survey-report-FINAL.pdf.

were unable to access them. People with vision or hearing disabilities discovered that their usual means of access didn't work. For example, many of the screen reader software tools used by people with low vision were not compatible with telehealth platforms [24]. A lack of closed captioning and sign language interpretation also caused difficulties for people who were Deaf or Hard of Hearing. These challenges remained on top of the difficulties of not having access to reliable caregivers due to a worker shortage caused by illness or quarantine. Thus, COVID-19 exposed deficiencies in the healthcare system that disproportionally affected those with disabilities.

Other access barriers related to emergencies and disasters have been reported. People who use wheelchairs for mobility require buildings that are physically accessible. During disasters, accessible disaster shelters become particularly important for these individuals as staying with family or friends may not be feasible if their homes are not accessible. However, although accessibility is a required by law, in reality disaster shelters may not be accessible. Disabled people have also been turned away from emergency shelters during disasters. For example, during Hurricane Irene (2011) a woman who used a wheelchair was turned away from an emergency shelter [25]. The gate for the ramp was locked and shelter staff didn't have the key. So the woman was forced to return home to potential danger from the storm. There were also reports of dangerously steep ramps, inaccessible shelter restrooms and cots, as well as a lack of American Sign Language (ASL) interpreters to provide accessible and effective communication [26]. The Center for Independence of the Disabled, NY (CIDNY) reported multiple communication access issues during Hurricane Irene, including paper evacuation instructions which were inaccessible to Blind people and New York City webpages, such as the online shelter locator tool, which were incompatible with screen reader software making the pages unusable for people with vision disabilities [27,28].

A 2018 report by the Partnership for Inclusive Disaster Strategies noted a number of access barriers to applying for FEMA post-disaster aid [8]. Barriers included challenges understanding the application process and completing the forms to receive individual disaster assistance as well as the lack of a place on the form to identify as a person with a disability who might have specific needs. For example, people who are Deaf or Hard of Hearing and whose primary communication mode or language is American Sign Language (ASL) reported that they were unable to understand the required forms. Since ASL is a different language than English, with different rules of grammar, among other things, being fluent in ASL does not mean automatic English reading fluency. In fact, many hearing people who are native English speakers also identified difficulty understanding the form. As well, people with low vision were unable to navigate through the form using their screen reader software. Others, with cognitive disabilities, intellectual disabilities and mental illness reported that they were also unable to understand the form and many lacked either computers skills or access to computers to be able to complete the forms online [8]. Physical access barriers to shelters during the disaster and inaccessible post disaster recovery services, among other obstacles, resulted in disparities. In turn this meant that it was harder for people with disabilities to weather the storm than people without disabilities.

Infringement of legal rights

While legal issues will be addressed in another chapter, it is important to note that civil rights violations may lead to negative health impacts or daily disruptions as well. For example, in testimony before a New York City Super Storm Sandy oversight committee, an attorney testified that individuals who lived in group homes (supervised homes in the community, often with four to six residents), who were usually able to come and go independently and control their own money, were evacuated to shelter sites intended for people who needed nursing home level care, based on false assumptions about residents' needs [29]. These individuals needed less intensive care than would be required by someone who lived in a nursing facility, who would likely have significant chronic illness and physical disabilities. Under the ADA people with disabilities are to be sheltered in the least restrictive environment or accommodated in the most integrated setting, based on their needs [12]. As the access and functional needs of this group were likely support-related, those needs could have been met in a mass care (public) shelter, rather than an environment with a higher level of care. The attorney testified that disaster shelter staff prohibited people with developmental disabilities from holding their own money while in the shelter and prevented them from leaving the shelter unaccompanied, even though they did so in their own homes. The attorney argued that these infringements were due to bias and false assumptions about disability [29]. Understanding that people with disabilities have different levels of need can help avoid these kinds of situations. Additionally, thoughtful inclusive planning would have identified the kinds of access and functional needs likely in the community, provided a better understanding about appropriate and respectful interactions with people with disabilities, and facilitated a focus on sheltering people with disabilities in the most integrated environment.

The COVID-19 pandemic has highlighted another issue requiring thoughtful planning and understanding of individual needs. For some people with disabilities, having a companion to provide support and/or aid communication is critical, particularly in situations like hospitalizations where critical information must be shared, and important care decisions made. Because of the concern about the spread of COVID-19, many hospitals have "no visitor" rules, which seem to include companions who provide support. The US Department of Health and Human Services and the California Department of Health have recognized the inequities inherent in these policies. The US Department of Health and Human Services Office for Civil Rights (OCR) received a number of complaints from people denied a companion for support while in the hospital. In one case a 73 year old Connecticut (CT) woman with aphasia (a condition that limits speech and understanding) wanted support in the hospital to help her effectively communicate and understand healthcare issues [30]. A hospital refused her request due to its COVID-19 "no visitor" policy. OCR worked with the parties to resolve the issue and the woman was permitted a companion. As part of the resolution, the State of CT issued an executive order establishing a state policy that people with disabilities would be permitted a companion for support in acute care settings. In California (CA) as part of their official Visitor Limitations Guidance for healthcare facilities, the CA Department of Health provided specifically for hospital visitors with disabilities to have companions for support [30–32].

Calls to action

The significant negative impacts of disasters on people with disabilities particularly since Hurricane Katrina (2005), have mobilized the disability community to strongly advocate for change. Many disability organizations, including the National Council on Disability, have issued calls to action and strongly encourage inclusive emergency planning [8,10,21]. The mission of the Partnership for Inclusive Disaster Strategies, a disability led organization, is focused on equal access to services before, during and after emergencies [33].

Centers for Independent Living (CILs), organizations developed and operated by disabled people and funded in part by the federal government, also advocate for inclusive planning and equal access [34]. The CILs provide independent living services to people with disabilities, including tools and resources to integrate people with disabilities into their communities based on a person-centered and person-directed philosophy. Many CILs are strong advocates for inclusive emergency management, lead inclusive emergency efforts, and provide disaster-related services to consumers. The Washington state CIL association, for example, leads the Coalition on Inclusive Emergency Planning (CIEP), a statewide disability group that provides situational awareness during disasters and writes after action reports (post-disaster assessments) for emergency management, community organizations, and the state CIL network [35].

Disability advocates have also tried to achieve inclusive emergency planning by instituting legal actions such as *Brooklyn Center for Independence of the Disabled (BCID), et al. v. Mayor Bloomberg, et al. 2013* [36]. In this case, the judge found that New York City's emergency plan failed to accommodate the emergency needs of people with disabilities. The parties' settlement called for a number of remedial actions to modify the New York City emergency plan to better meet the needs of people with disabilities [36].

Impacts of emergencies on other populations with access and functional needs

Although the focus of this book is on people with disabilities, it must be acknowledged that other populations with access and functional needs experience similar negative impacts and disparities. Research confirms that while traditional emergency plans assumed children to have the same emergency needs as adults, in fact they have specialized needs based on age, immature physiology, and level of social and emotional development. Lack of understanding of children's disaster needs and lack of attention to meeting those needs has resulted in negative impacts on children's mental health and development post disaster [37]. The National Commission on Children in Disasters and child welfare organizations have issued strong statements expressing concern about the impacts of disasters on children, including children with disabilities, because of poor advance planning [38,39]. These child advocates have issued strong calls to action.

Older adults have experienced negative impacts as well and studies have shown that they are more likely to die in a disaster than younger people. This was true during both Hurricane Katrina in 2005 and Superstorm Sandy in 2012, where almost half or more of the deaths were among older adults [40]. The magnitude of this impact is described in the statement in Fig. 2. Social isolation and living alone, common among older adults, make them particularly vulnerable, because they may be without help or able to help themselves when an emergency happens.

Other researchers have examined discrimination against other groups with potential access and functional needs, including the challenges faced by minority group members and those with limited incomes. An Amnesty International Report reveals not only post-hurricane housing discrimination and limited access to health care for African Americans with limited incomes, but also highlights police brutality to African Americans during Hurricane Katrina [41]. The legal actions related to these events are on-going [42]. An excerpt from the NAACP website found in Fig. 3 reflects the potential disaster vulnerability of communities of color.

Environmental disasters like Hurricane Katrina are not the only types of disasters that disproportionately impact BIPOC (Black, Indigenous, People of Color) populations; the COVID-19 pandemic

"In Louisiana, about 71 percent of those who died as a result of Hurricane Katrina were older than 60, and nearly half were older than 75, according to a 2006 federal report. About two weeks after Hurricane Sandy hit in 2012, the New York Times reported that close to half of those who died in the storm were 65 or older. Many of these elderly victims drowned at home; others died from storm-related injuries, hypothermia and other causes."

FIG. 2

Impact of Hurricanes Sandy and Katrina on older adults.

Excerpted from Parry W. Why disasters like Sandy hit the elderly hard. Live Science; 2013. Available from: https://www.livescience. com/27752-natural-disasters-hit-elderly-hard.html.

Environmental disasters cause widespread disruptions to daily life from which a community, individual, or organization could not recover without outside assistance. These kinds of threats are not randomly distributed. Communities of color and other frontline communities tend to live in the most at-risk environments and are more vulnerable to the negative impacts of these kinds of events due to a range of preexisting factors. [42]

FIG. 3

NAACP, *In the Eye of the Storm.*

Excerpted from Steicheny L, Patterson J, Taylore K. NAACP Environmental & Climate Justice Program in the eye in the eye of the storm a people's guide to transforming crisis and advancing equity in the disaster continuum; 2018, Rev'd 2021 by Abdul-Rahman, D. Available from: https://naacp.org/resources/eye-storm-peoples-guide-transforming-crisis-advancing-equity-disaster-continuum.

has had an analogous effect [43]. For example, a study conducted by amfAR, The Foundation for AIDS Research, found COVID-19 cases and deaths increased in counties as the proportion of Black residents rose [44]. Counties with predominantly Black residents accounted for a higher proportion of COVID-19 cases and deaths than were found in counties with fewer Black residents. After analysis, the study concluded that these disparities could not be explained solely by the presence of underlying conditions. In a study describing the impact of COVID-19 on communities of color, Millett noted a similar finding of disproportionate impact due to H1N1 influenza in 2009–2010, a decade earlier than the COVID-19 pandemic [43]. In 2020 in order to address race discrimination during emergencies, five federal agencies issued joint guidance to recipients of federal financial assistance for emergency management. This group of agencies consisted of the United States Departments of Justice (DOJ), Homeland Security (DHS), Housing and Urban Development (HUD), Health and Human Services (HHS), and Transportation (DOT). The guidance reiterates that unlawful discrimination during emergencies violates the Civil Rights Act of 1964, Title VI and other laws [45,46]. Based partly on lessons learned during and after Hurricane Katrina and on the Disaster Recovery Framework in addition to the Civil Rights Act, this document seeks to ensure that disaster management actions neither intentionally nor unintentionally disadvantage individuals or groups because of their race, color, or national origin (including limited English proficiency), religion, sex or disability. The guidance emphasizes the importance of providing equal access to services and resources to meet the needs of the whole community.

Despite federal guidance and multiple calls to action by various groups, minority, disability and other access and functional need populations (such as those with limited financial resources) may face additional barriers to inclusive emergency planning. Foster-Bey writes that race, ethnicity, and family income predict civic engagement, with individuals with low incomes and minority group members engaging less frequently in civic participation activities than people who are Caucasian and/or have higher incomes [47]. Outreach to the BIPOC community, similar to outreach to the disability community, should be tailored to community needs and culture. Additionally, many people identify with multiple different populations that have access and functional needs. A person with one or more disabilities can also be Asian American or have limited English language proficiency. They can be part of a family with limited financial resources that lives in a low income neighborhood. For this reason, it is important to consider "intersectionality" or the overlap between the identification of a single individual with multiple, different groups [48]. The overlap of these multiple forms of identity may compound discrimination and bias, leading to an even greater vulnerability to the negative impacts of emergencies, disasters and pandemics (see, e.g., Ref. [49]).

Systems issues: The importance of having a seat at the table

In order to effectively assure that the needs of the whole community, including people with disabilities and those with multiple vulnerabilities, are met, the causes of these negative impacts need to be assessed and addressed. Possible causes related to individuals with disabilities include: (1) lack of personal preparedness; (2) nature of a disability diagnosis; and (3) lack of knowledge about how the American emergency management system works. Potential causes related to emergency managers and public health preparedness planners include: (1) Knowledge about and experience with people with disabilities; (2) attitudes toward inclusive emergency planning and stakeholder engagement; and (3) knowledge gaps about ADA implementation.

Personal preparedness

Lack of personal preparedness does not necessarily provide the sole answer, because it is unclear whether people with disabilities are more or less prepared for emergencies than people without disabilities. Personal preparedness is thought to lead to greater resilience on the part of people with disabilities as well as those without disabilities. Personal responsibility for individual and household preparedness is emphasized by the US government, for example in the ReadyGov website, as a way for individuals to mitigate disaster impacts [50]. Preparedness data on individuals with disabilities is mixed, with some studies indicating that people with disabilities are as underprepared as the general population and others showing that disabled people (or people with particular kinds of disabilities) are less prepared than people without disabilities [51–53]. For example, at least one study found that people with disabilities were less well-prepared than people without disabilities generally but were more likely to have a 3 days supply of medication [52]. Another study found that having a disability that limited activities or a mental health conditions increased the likelihood of being unprepared, more than having a disability and using adaptive equipment or not having a disability [51]. And still another study found that "medically vulnerable" adults (including adults with disabilities) were less likely to be prepared than people without medical vulnerabilities [54]. Gershon and colleagues [53] provide other examples of the inconsistencies.

But a FEMA study assessing a "preparedness profile" consisting of a number of preparedness factors found minimal differences between people with and without disabilities [55]. The factors examined, and results of the study appear in Fig. 4. For example, people with disabilities, caregivers for

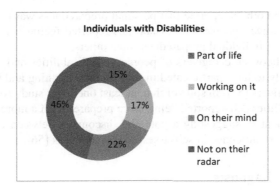

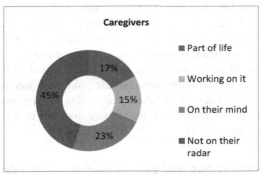

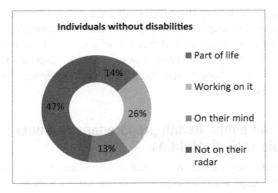

FIG. 4

Preparedness profiles of people with disabilities, caregivers, and people without disabilities.

Adapted from The U.S. Department of Homeland Security FEMA. Preparedness in America research insights to increase individual, organizational, and community action; 2014. Available from: https://www.ready.gov/sites/default/files/2021-03/Preparedness%20 in%20America_August-2014.pdf.

people with disabilities, and others reported that personal preparedness was a "part of life" in roughly the same percentages. However, people with disabilities reported feeling vulnerable and having less confidence in the efficacy of individual preparedness than others.

The major difference between caregivers of people with disabilities and non-caregivers was that caregivers were more likely to have participated in preparedness training and have a household emergency plan than non-caregivers [55]. In spite of that, at least one other study found that parent caregivers of children with disabilities self-reported being more prepared than a more objective measurement of their preparedness indicated, suggesting a potential disconnect between objective and subjective measures and potential overconfidence in self-assessed preparedness [56].

Nature of a disability diagnosis

The nature of a person's disability diagnosis and their individual vulnerability is another factor that could place them at greater risk. At least one group of researchers found that the nature of certain disabilities made it more difficult to take preventative actions, such as mask wearing [57]. For example, some individuals with autism spectrum disorder have sensitivities to particular sensations which can lead to behavioral disruptions. Mask wearing is one such sensation that, while uncomfortable for many people, may be magnified in a person with autism to the point that they are unable to wear a mask. However, a study from the United Kingdom found that having an impairment in and of itself didn't account for the vulnerability to environmental disasters of disabled people, and argues that inherent vulnerability should not be assumed [58].

Knowledge about the emergency and public health preparedness systems

Another factor may be that people with disabilities (like many Americans) lack understanding about how the American emergency management and public health preparedness systems function, and therefore may have unrealistic expectations and feel less of a need to fully prepare. In an online survey of emergency managers and public health preparedness planners, respondents expressed concern that the disability community had knowledge gaps about how the emergency system works and the roles and responsibilities within the system [59].

Emergency manager and public health preparedness planner knowledge and experience with people with disabilities

Systemic issues within emergency management, rather than limited personal preparedness, disability diagnosis, and/or knowledge gaps about emergency management systems play a major role in causing the negative impacts of disasters on people with disabilities. While the goal of emergency management is to provide protection from the physical, psychological, economic, and social harms caused by natural or human caused disasters and emergencies, people with disabilities and others with access and functional needs have been repeatedly and disproportionally harmed by disasters [60,61]. Historically, the United States has neglected the disaster needs of potentially vulnerable populations such as those with access and functional needs (AFN) by excluding them from the local emergency planning process and omitting their likely emergency needs from local plans [62,63]. By not recognizing that everyone cannot walk, run, talk, drive, and quickly follow directions or may not have enough knowledge, ability and resources to care for themselves, emergency management has effectively excluded many people from the protections offered by their community when disaster strikes. Furthermore, it excludes the same people from the benefits of engaging in the processes guiding planning, mitigation, response, and recovery. Commentators report a concomitant knowledge gap among emergency planners that appears to be widespread as evidenced in multiple small and large disasters since at least the September 11, 2001 terror attacks [8,64]. Increasingly, federal, state, and local emergency management and public health agencies have emphasized filling knowledge and planning gaps via collaborative and inclusive emergency planning. White describes how an effective collaborative process can work. There is an excerpt from his work in Fig. 5 [65].

In this vein, a number of disability-related trainings and resources have been developed and made available for emergency managers and public health preparedness planners (see, e.g., Ref. [66]). For example, FEMA's Emergency Management Institute has several online and in-person courses related to meeting the needs of people with disabilities and others with access and functional needs [67]. FEMA's *Developing and Maintaining Emergency Operations Plans Comprehensive Preparedness Guide (CPG) 101*, the United States' foundational disaster planning guide, also emphasizes the importance of meeting access and functional needs issues [68]. The TRAIN Learning Network (TRAIN) for public health education offers additional online courses, and the Pacific ADA Center also has an extensive webinar series on the topic [69,70]. The US Department of Health and Human Services Office of the Assistant Secretary for Preparedness and Response (ASPR) has online information resources [71] and an online access and functional needs training posted to the TRAIN network [72]. There are also healthcare and public health emergency preparedness and response capabilities, developed by ASPR and the CDC, respectively [73,74], some of which address access and functional needs populations. These resources set out the knowledge and capabilities required to begin to incorporate people with disabilities into public health preparedness processes. However, by themselves, these resources are not sufficient to assure that the integration of the emergency needs of people with disabilities occurs. Rather, additional factors need to be addressed in order to achieve that goal.

Whole community inclusive emergency planning and public engagement

One factor related to the gap in addressing these needs may be emergency manager and public health preparedness planner attitudes toward "whole community" or inclusive emergency planning and response. By definition, inclusive emergency management involves public engagement in the process. When engaged, stakeholders bring with them unique expertise via their own lived experience. In the

> "...[A] county emergency manager who recruits knowledgeable consumers to help inform disability-friendly policies for evacuation and shelter can drastically affect the consumers' quality of life throughout the disaster cycle. If people with disabilities are at the table and actively involved in planning for possible disaster scenarios and emergencies, they are more likely to have an emergency management plan in place to be evacuated versus being triaged or forgotten."

FIG. 5

Including people with disabilities.

Excerpted from p. 168, White GW. Nobody left behind: disaster preparedness and public health response for people with disabilities. In: Drum CE, Krahn GL, and Bersani H, editors. Disability and public health. Washington, DC: American Public Health Association; 2009. p. 163–82.

book *Nothing About Us Without Us*, Charlton expresses the belief that people with disabilities are the main experts about their own lives and best know their own needs [75]. A well-known disability rights slogan, "nothing about us without us," which is thought to have originated as a political motto in Central Europe, especially Poland, during the 1500s [76], refers to the idea that no disability-related policy should be created without the full, direct participation of the people who will be affected by the policy [75]. If people with disabilities are the main experts on their own lives, it then follows that they should be involved in developing emergency management and public health preparedness policies and practices that impact them. In other words, disabled people, and other populations with access and functional needs, should have a seat at the emergency planning table.

Evidence shows that emergency planners approach inclusive planning in various ways. Some research indicates that emergency managers see inclusive emergency planning and response as a positive opportunity to engage the public (see, e.g., Ref. [77]). However, Sievers reports a study indicating that while emergency managers reportedly valued civic engagement, they did not practice it. Instead, they engaged in emergency planning only with consultants or a small local government planning team [78]. Another study found that few emergency managers made targeted outreach to vulnerable populations and that emergency manager lack of disability awareness has persisted [79]. This finding is consistent with results of an emergency and public health preparedness planner survey by the Association of University Centers on Disabilities' (AUCD) [59]. The study results indicate that while acknowledging the importance of input from the disability community, only 25% of the emergency and public health preparedness planners surveyed actually engaged with the disability community. Lack of awareness of how to make outreach to the disability community may contribute to this situation.

Sobelson, et al. argue that civic engagement is not novel to public health agencies as they routinely engage in these activities [80]. Schoch-Spana and colleagues concur with this, but also note that while local public health agencies have deepened whole community engagement over time, stagnation has occurred recently due to budgetary and staff constraints [81]. Ramsbottom, and colleagues, on the other hand, assert that public health emergency plans generally remain top down, with limited public consultation except by experts [82]. A number of factors potentially contribute to this lack of inclusion. Like emergency managers, public health preparedness planners may face knowledge gaps and may not recognize the benefit of involving the whole community. Surveys of public health departments show that they generally don't rate community collaborations highly. According to one study, the public health departments surveyed rated these collaborations between "fair" and "good" [83]. Sobelson and colleagues report public health preparedness planner concerns that community residents will have unrealistic expectations of the community's emergency capabilities which could bias public health preparedness planner perspectives [80]. However, studies show that public health planners' perception of public beliefs and preparedness do not necessarily match what the public actually thinks [84]. For example, members of the public think that they are more informed than public officials think they are. Also, public expectations seem to differ between demographic groups and appear to be somewhat the product of their understanding of the capabilities of the emergency response [85]. Expectations can also be created, such as during Hurricane Isabel when the city of Richmond delivered ice to its citizens. The ice did not add much to the recovery activities, however, people still expected ice to be delivered in the event of a disaster. Thus, expectations develop in a complex manner.

In addition to issues related to stakeholder engagement and expectations, there is a question of knowledge of how to implement the ADA mandates related to emergencies, disasters and pandemics [11]. Gershon and colleagues surveyed 61 local emergency managers in FEMA Region 9, which includes four states (Arizona, California, Hawaii, Nevada), five territories (Guam, American Samoa, Commonwealth of Northern Mariana Islands, Republic of Marshall Islands, Federated States of Micronesia) and 150 tribal nations [86]. The study results shown in Fig. 6 include emergency manager self-report related to knowledge gaps about disability demographics and gaps in emergency plans and practices.

The emergency managers surveyed reported a desire for more staff training on the ADA and better outreach to disabled people and others with access and functional needs.

The results of the AUCD survey of emergency managers and public health professionals described above indicate that while emergency and public health professionals recognize the importance of engaging with people with disabilities, only a minority have actually incorporated engagement into their planning process [59]. In that survey the respondents reported gaps in their understanding of the emergency rights of disabled people and how to include them in inclusive emergency planning. Another key finding was the perception that a major challenge was a lack of shared understanding among the disability community and planners as to emergency event responsibilities and experiences [59]. Funding and staff capacity barriers were also reported as in Gershon and colleagues' survey [86].

Other surveys, such as one completed by the National Association of County and City Health Officials (NACCHO), look at local public health departments' knowledge and awareness of disabled people in

Percent of Respondents (N=61)	Emergency managers reported:
57%	No access to disability and access and functional need population demographics.
63%	Emergency plan addressed disability-related needs. • 43% reported input from disability organizations or disabled people in plan development. • 41% reported specific operating procedures related to people with disabilities.
34%	Emergency warnings for people with disabilities tested for people with disabilities
52%	Pre-identified accessible transport
69%	Pre-identified accessible shelters • 70% reported ramps • 38% reported wheelchair accessible bathrooms • 56% reported barrier free areas • 18% reported use of alternative communication modalities (e.g. Braille) • 46% reported power generators • 49% reported refrigerators for medication • 39% reported policies to permit personal assistance providers to come to the shelter to help clients • 62% reported policies to permit service animals • 18% reported sheltering people with disabilities in medical shelters absent medical need

FIG. 6

Emergency manager self-report survey.

Excerpted from Gershon RR, Muska MA, Zhi Q, Kraus LE. Are local offices of emergency management prepared for people with disabilities? Results from the FEMA Region 9 Survey. J Emerg Manag. 2021;19(1):7–20. https://doi.org/10.5055/jem.0506 PMID: 33735431.

"[Emergency managers] have hesitated to undertake inclusive preparedness initiatives due to lingering confusion of the type and nature of services that are legally required under the ADA and frustration over the perception that federal emergency preparedness guidelines set unrealistic and unattainable standards of compliance."

FIG. 7

From Sherry and Harkin.

From Sherry N, Harkins AM. Leveling the emergency preparedness playing field. J Emerg Manag 2011;9(6):11–6. https://doi. org/10.5055/jem.2011.0075.

their community [87]. In 2018, 46% of responding health departments reported being "aware or very aware" of the number of people with disabilities in their jurisdiction and only 10% of health departments felt that they were "not at all aware" [87]. More than half of health departments surveyed described themselves as being "knowledgeable or very knowledgeable" about disability accommodations and understanding disparities experienced by this population [87]. More than 85% of local health departments reported having an inclusive emergency preparedness program that engages with disability service provider organizations. A majority of local health departments reported offering disability-related training.

Despite the large numbers of local public health departments that incorporate people with disabilities into their activities, the NACCHO survey noted that many local health departments also had needs related to the inclusion of people with disabilities. These needs included funding accommodations, having enough time and resources for inclusion, and lack of training. Like their colleagues in emergency management, public health professionals indicated gaps in awareness about how to address disability-related needs in emergency planning [87]. NACCHO found that local health departments also displayed a lack of awareness of the importance of inclusive programs even though, similar to their emergency management colleagues, they understood the importance of engaging the disability community. Local health departments also indicated a reluctance of people to identify themselves as disabled because of concerns about negative consequences.

The findings related to emergency management and public health professionals' ongoing confusion about how to effectively engage people with disabilities and include them in planning support the conclusions of Sherry and Harkins from about 10 years previous [62]. They wrote the following, found in Fig. 7, about emergency manager confusion about the ADA.

While there may be on-going confusion, the survey results seem to identify some opportunities as well. For example, if public health and emergency management agencies partnered with each other and local government ADA coordinators (the latter required for state and local governments with more than 50 employees), then some of these gaps might be closed [88].

Conclusion

Too often the negative impacts of health inequities related to emergencies continue despite attempts to reverse them through mechanisms such as civil rights laws, education, and advocacy. Several factors contribute to this situation. First, emergency plans may assume that all of us can talk, hear, drive, walk, run and quickly follow directions without adaptation or modification. Second, local emergency, public health preparedness planners and health care providers who work with these populations and participate in planning and response may lack the knowledge required to successfully address disability-related needs. Finally, people with disabilities and disability organizations often lack a seat at the local emergency planning table and may be uncertain about how to become involved in order to advocate effectively for disability inclusion. A seat at the table, either literally at the emergency planning table or via strong input into the planning process in other ways, would serve to decrease these negative impacts and disparities, support civil rights, and help build both individual and community resilience.

Questions for thought

(1) Choose one population with access and functional needs, whether or not mentioned in this chapter. Read one article about the population's negative or positive experiences with emergencies and disasters and identify one or two key issues from the article. Next, think about how a local emergency plan might address barriers or build on strengths that you have identified from the article. Then draft the text you would add to the local plan to address the issue(s).

(2) Too often Americans are unprepared or underprepared for emergencies and disasters. What do you think are the main reasons? Does the COVID-19 pandemic and the strong feelings about COVID-19 vaccination provide any clues to the reasons for lack of preparedness?

(3) Think about the population you chose for Question 1. Now design two social media messages as part of a campaign to increase preparedness among this population.

References

[1] Ndugga N, Artiga S. Disparities in health and health care: 5 key questions and answers, 2021. [cited 2022 Jan 2]. Available from: https://www.kff.org/racial-equity-and-health-policy/issue-brief/disparities-in-health-and-health-care-5-key-question-and-answers/.

[2] Centers for Disease Control and Prevention, National Center on Birth Defects and Developmental Disabilities, Division of Human Development and Disability, Disability and Health Program. Disability and health state chartbook, 2006 profiles of health for adults with disabilities. Atlanta, GA: Centers for Disease Control and Prevention; 2006. Available from: https://stacks.cdc.gov/view/cdc/11416.

[3] United Nations, Department of Social and Economic Affairs. COVID-19 outbreak and persons with disabilities, 2021. Available from: https://www.un.org/development/desa/disabilities/covid-19.html.

[4] Krahn GL, Walker DK, Correa-De-Araujo R. Persons with disabilities as an unrecognized health disparity population. Am J Public Health 2015;105(S2):S198–206. https://doi.org/10.2105/AJPH.2014.302182.

[5] Testimony of Paul Timmons. Disaster preparedness and response: The special needs of older Americans: Hearings before the Special Committee on Aging, Senate, 115 Cong, 2017. Available from: https://www.aging.senate.gov/hearings/disaster-preparedness-and-response-the-special-needs-of-older-americans.

[6] Missildine K, Varnell G, Williams J, Grover KH, Ballard N, Stanley-Hermanns M. Comfort in the eye of the storm: a survey of evacuees with special medical needs. J Emerg Nurs 2009;35(6):515–20. https://doi.org/10.1016/j.jen.2009.07.007.

[7] Kishore N, Marqués D, Mahmud A, Kiang MV, Rodriguez I, Fuller A, et al. Mortality in Puerto Rico after Hurricane Maria. N Engl J Med 2018;379(2):162–70. https://doi.org/10.1056/NEJMsa1803972.

[8] Roth M, Kailes JI, Marshall M. Getting it wrong: An indictment with a blueprint for getting it right disability rights, obligations and responsibilities before, during and after disasters (edition 1), 2018 May. Available from: http://www.disasterstrategies.org/index.php/news/partnership-releases-2017-2018-after-action-report.

[9] Perry D. No, we should not involuntarily commit the homeless during hurricanes. Pacific Standard, 2017 Sep 15. Available from: https://psmag.com/social-justice/we-should-not-involuntarily-commit-the-homeless-during-hurricanes.

[10] National Council on Disability. Preserving our freedom ending institutionalization for people with disabilities during and after disasters, 2019. Available from: https://ncd.gov/publications/2019/preserving-our-freedom.

[11] Americans with Disabilities Act. 42 U.S. Code § 12101 et seq; 1990.

[12] U.S. Department of Health and Human Services, Office for Civil Rights. Serving people with disabilities in the most integrated setting: Community living and Olmstead. Available from: https://www.hhs.gov/civil-rights/for-individuals/special-topics/community-living-and-olmstead/index.html.

[13] Centers for Medicare and Medicaid Services. Long term services and supports rebalancing toolkit, 2020 Nov. Available from: https://www.medicaid.gov/medicaid/long-term-services-supports/downloads/ltss-rebalancing-toolkit.pdf.

[14] Gleason J, Ross W, Fossi A, Blonsky H, Tobias J, Stephens M. The devastating impact of Covid-19 on individuals with intellectual disabilities in the United States. NEJM Catal Innov Care Deliv 2021;2(2). https://doi.org/10.1056/CAT.21.0051.

[15] Jenkins A. Why is COVID-19's death toll so high for people with developmental disabilities? Oregon Public Broadcasting (OPB); 2020. Available from: https://www.opb.org/news/article/covid-19-death-toll-developmental-disability/.

[16] Fair Health. Risk factors for COVID-19 mortality among privately insured patients: A claims data analysis, 2020. Available from: https://www.fairhealth.org/publications/whitepapers.

[17] Wadman M. COVID-19 is 10 times deadlier for people with Down syndrome, raising calls for early vaccination. Science 2020;, 121520. Available from: https://www.science.org/content/article/covid-19-10-times-deadlier-people-down-syndrome-raising-calls-early-vaccination.

[18] Hippisley-Cox J, Coupland CA, Mehta N, Keogh RH, Diaz-Ordaz K, Khunti K, et al. Risk prediction of covid-19 related death and hospital admission in adults after Covid-19 vaccination: national prospective cohort study. BMJ 2021;374. https://doi.org/10.1136/bmj.n2244.

[19] Centers for Disease Control and Prevention. People with certain medical conditions. Centers for Disease Control and Prevention; 2021. [cited 2021 Dec 21]. Available from: https://www.cdc.gov/coronavirus/2019-ncov/need-extra-precautions/people-with-medical-conditions.html.

[20] The COVID tracking project at The Atlantic long-term-care COVID tracker. The Atlantic, 2021. Available from: https://covidtracking.com/nursing-homes-long-term-care-facilities.

[21] National Council on Disability. Effective emergency management: Making improvements for communities and people with disabilities, 2009. Retrieved from: https://ncd.gov/publications/2009/Aug122009.

[22] Disability Rights Texas. The forgotten faces of winter storm Uri. Disability Rights Texas; 2021. [cited 2021 Dec 21]. Available from: https://media.disabilityrightstx.org/wp-content/uploads/2021/04/06100917/apr-5-2021-DRTX-winter-survey-report-FINAL.pdf.

[23] Shalby C. Power outages leave those with disabilities especially vulnerable. Help remains a work in progress. Los Angeles Times 2019. 10/25/2019; Available from: https://www.latimes.com/california/story/2019-10-25/problems-disabled-help-power-outages.

[24] National Institute for Health Care Management (NIHCM) Foundation. Disability, health equity & COVID-19 coronavirus. NICHM Foundation; 2021. 09/21/21 Updated 10/14/21. Available from: https://nihcm.org/publications/disability-health-equity.

[25] Anon. Brooklyn Center for Independence of the Disabled (BCID), et al. v. Bloomberg, No. 11 Civ. 6690. Class Action Complaint for Discrimination; Injunctive and Declaratory Relief. (S.D.N.Y. September. 26, 2011)., 2011. Available from: https://dralegal.org/case/brooklyn-center-for-independence-of-the-disabled-bcid-et-al-v-mayor-bloomberg-et-al/.

[26] The Council of the City of New York General Welfare Committee. Briefing paper of the human services & government affairs division for oversight hearing: Emergency planning and management during and after Hurricane Sandy: Shelter management, 2013. 022013. Available from: https://studylib.net/doc/7390466/general-welfare-committee-staff--andrea-vazquez--counsel.

[27] Center for Independence of the Disabled, NY. CIDNY emergency preparedness presentation 2015, 2015. Available from: https://www.cidny.org/wp-content/uploads/2017/07/CIDNY-Emergency-Preparedness-Presentation-2015.pdf.

[28] Anon. Declaration of Susan Dooha, in Brooklyn Center for Independence of the Disabled (BCID), et al. v. Bloomberg. 08302012. Available from: https://www.cidny.org/wp-content/uploads/2017/07/037_Dooha_Decl.pdf.

[29] Testimony of Shelley Weizman. Oversight: Emergency planning and management during and after the storm: Emergency preparedness and response at the city's healthcare facilities, 2013. Available from: http://mobilizationforjustice.org/wp-content/uploads/Disaster-Planning-Testimony-1-24-13.pdf.

[30] U.S. Department of Health and Human Services. OCR resolves complaints after State of Connecticut and private hospital safeguard the rights of persons with disabilities to have reasonable access to support persons in hospital settings during COVID-19. U.S. Department of Health and Human Services; 2020. [cited 2022 Jan 2]. Available from: https://public3.pagefreezer.com/content/HHS.gov/31-12-2020T08:51/https://www.hhs.gov/about/news/2020/06/09/ocr-resolves-complaints-after-state-connecticut-private-hospital-safeguard-rights-persons.html.

[31] Disability Rights and Education Fund. Know your rights: Hospital visitation. Disability Rights Education Fund; 2020. [cited 2022 Jan 2]. Available from: https://dredf.org/know-your-rights-hospital-visitation/.

[32] State of California—Health and Human Services Agency California Department of Public Health. Visitor limitations guidance, 2021 Jun. Available from: https://www.cdph.ca.gov/Programs/CHCQ/LCP/Pages/AFL-20-38.aspx.

[33] The Partnership for Inclusive Disaster Strategies. Mission. The Partnership for Inclusive Disaster Strategies; 2022. [cited 2022 Jan 7]. Available from: https://disasterstrategies.org/mission/.

[34] Administration for Community Living. Centers for independent living. U.S. Department of Health and Human Services Administration on Community Living; 2021. [cited 2021 Dec 21]. Available from: https://acl.gov/programs/aging-and-disability-networks/centers-independent-living.

[35] Washington State Independent Living Council. Coalition on Inclusive Emergency Planning (CIEP). Washington State Independent Living Council; 2020. [cited 2021 Dec 21]. Available from: https://www.wasilc.org/coalition-on-inclusive-emergency-planning.

[36] Anon. Brooklyn Center for Independence of the Disabled (BCID), et al. v. Bloomberg, No. 11 Civ. 6690 (JMF) Stipulation of Settlement and Remedial Order. (S.D.N.Y. Nov. 7, 2013)., 2013. Available from: https://dralegal.org/case/brooklyn-center-for-independence-of-the-disabled-bcid-et-al-v-mayor-bloomberg-et-al/.

[37] Dziuban EJ, Peacock G, Frogel M. A child's health is the public's health: progress and gaps in addressing pediatric needs in public health emergencies. Am J Public Health 2017;107(S2):S134–7. https://doi.org/10.2105/AJPH.2017.303950.

[38] National Commission on Children in Disasters. National commission on children in disasters 2010 report to the president and congress [AHRQ Publication No. 10-M037 October 2010]; 2010, ISBN:978-1-58763-401-7. Available from: https://archive.ahrq.gov/prep/nccdreport/nccdreport.pdf.

[39] Save the Children. Still at risk: U.S. children 10 Years after Hurricane Katrina 2015 national report card on protecting children in disasters, 2015. Available from: https://secure.savethechildren.org/atf/cf/%7B9def2ebe-10ae-432e-9bd0df91d2eba74a%7D/DISASTERREPORT_2015.PDF?v=5.

[40] Parry W. Why disasters like Sandy hit the elderly hard. Live Science; 2013. Available from: https://www.livescience.com/27752-natural-disasters-hit-elderly-hard.html.

[41] Amnesty International. Un-natural disaster human rights in the Gulf coast 2010, 2010. Available from: https://www.amnestyusa.org/katrina-six-years-on-criminal-justiceand-the-nopd/.

[42] Steicheny L, Patterson J, Taylore K. NAACP Environmental & Climate Justice Program in the eye in the eye of the storm a people's guide to transforming crisis and advancing equity in the disaster continuum, 2018. Rev'd 2021 by Abdul-Rahman, D. Available from: https://naacp.org/resources/eye-storm-peoples-guide-transforming-crisis-advancing-equity-disaster-continuum.

[43] Millett G. COVID-19, communities of color, & social determinants of health Presentation to Mental Health America. Available from: https://mhanational.org/bipoc-communities-and-covid-19.

[44] amfAR. The Foundation for AIDS Research, amfAR study shows disproportionate impact of COVID-19 on Black Americans, 2020 May. Available from: https://www.amfar.org/amfAR-Study-Shows-Disproportionate-Impact-of-COVID-19-on-Black-Americans/.

[45] Civil Rights Act of 1964, Title VI. 42 U.S.C. § 2000d Et Seq.

[46] The United States Departments of Justice (DOJ), Homeland Security (DHS), Housing and Urban Development (HUD), Health and Human Services (HHS), Transportation (DOT). Guidance to state and local governments and other federally assisted recipients engaged in emergency preparedness, response, mitigation, and recovery activities on compliance with Title VI of the Civil Rights Act of 1964, 2020. Available from: https://www.justice.gov/crt/file/885401/download.

[47] Foster-Bey J. Do race, ethnicity, citizenship and socio-economic status determine civic-engagement? In: CIRCLE working paper# 62. Center for Information and Research on Civic Learning and Engagement (CIRCLE); 2008 Dec. Available from https://circle.tufts.edu/sites/default/files/2019-12/WP62_RaceEthnicityCitizenshiSocioeconomicStatusandCivicEngagement__2008.pdf.

[48] Kuran CHA, Morsut C, Kruke BI, Krüger M, Segnestam L, Orru K, Nævestad TO, Airola M, Keränen J, Gabel F, Hansson S. Vulnerability and vulnerable groups from an intersectionality perspective. Int J Disaster Risk Reduct 2020;50:101826. https://doi.org/10.1016/j.ijdrr.2020.101826.

[49] Elisala N, Turagabeci A, Mohammadnezhad M, Mangum T. Exploring persons with disabilities preparedness, perceptions and experiences of disasters in Tuvalu. PLoS ONE 2020;15(10), e0241180.

[50] ReadyGov. Make a plan. Available from: https.www.ready.gov/plan (05/18/2022).

[51] Smith DL, Notaro SJ. Personal emergency preparedness for people with disabilities from the 2006-2007 Behavioral Risk Factor Surveillance System. Disabil Health J 2009;2(2):86–94. https://doi.org/10.1016/j.dhjo.2009.01.001.

[52] Bethel JW, Foreman AN, Burke SC. Disaster preparedness among medically vulnerable populations. Am J Prev Med 2011;40(2):139–43. https://doi.org/10.1016/j.amepre.2010.10.020.

[53] Gershon RR, Kraus LE, Raveis VH, Sherman MF, Kailes JI. Emergency preparedness in a sample of persons with disabilities. Am J Disaster Med 2013;8(1):35–47. https://doi.org/10.5055/ajdm.2013.0109.

[54] Barbato D, Bryie L, Carlisle CM, Doroodchi P, Dowbiggin P, Huber LB. Chronically unprepared: emergency preparedness status among US medically vulnerable populations. J Public Health 2021;1–9. https://doi.org/10.1007/s10389-021-01487-0.

[55] The U.S. Department of Homeland Security FEMA. Preparedness in America research insights to increase individual, organizational, and community action, 2014. Available from: https://www.ready.gov/sites/default/files/2021-03/Preparedness%20in%20America_August-2014.pdf.

[56] Wolf-Fordham S, Curtin C, Maslin M, Bandini L, Hamad CD. Emergency preparedness of families of children with developmental disabilities: what public health and safety emergency planners need to know. J Emerg Manag 2015;13(1):7. https://doi.org/10.5055/jem.2015.0213.

[57] Boyle CA, Fox MH, Havercamp SM, Zubler J. The public health response to the COVID-19 pandemic for people with disabilities. Disabil Health J 2020;13(3):100943. https://doi.org/10.1016/j.dhjo.2020.100943.

[58] Connon IL, Hall E. 'It's not about having a back-up plan; it's always being in back-up mode': rethinking the relationship between disability and vulnerability to extreme weather. Geoforum 2021;126:277–89. https://doi.org/10.1016/j.geoforum.2021.08.008.

[59] Association of University Centers on Disabilities (AUCD). What do local emergency preparedness staff think about disability inclusion? 2019. Available from: https://nationalcenterdph.org/our-focus-areas/emergency-preparedness/prepared4all/prepared4all-surveys-local-emergency-managers-public-health-preparedness-staff/.

[60] Nelson C, Lurie N, Wasserman J, Zakowski S. Conceptualizing and defining public health emergency preparedness. Am J Public Health 2007;97(Supplement 1):S9–11.

[61] Nicholson WC. Emergency planning and potential liabilities for state and local governments. State Local Gov Rev 2007;39(1):44–56. https://doi.org/10.1177/0160323X0703900105.

[62] Sherry N, Harkins AM. Leveling the emergency preparedness playing field. J Emerg Manag 2011;9(6):11–6. https://doi.org/10.5055/jem.2011.0075.

[63] Kailes JI, Enders A. Moving beyond "special needs" a function-based framework for emergency management and planning. J Disabil Policy Stud 2007;17(4):230–7. https://doi.org/10.1177/10442073070170040601.

[64] Engelman A, Ivey SL, Tseng W, Dahrouge D, Brune J, Neuhauser L. Responding to the deaf in disasters: establishing the need for systematic training for state-level emergency management agencies and community organizations. BMC Health Serv Res 2013;13(1):84.

[65] White GW. Nobody left behind: disaster preparedness and public health response for people with disabilities. In: Drum CE, Krahn GL, Bersani H, editors. Disability and public health. Washington, DC: American Public Health Association; 2009. p. 163–82.

[66] Wolf-Fordham SB, Twyman JS, Hamad CD. Educating first responders to provide emergency services to individuals with disabilities. Disaster Med Public Health Prep 2014;8(6):533–40. https://doi.org/10.1017/dmp.2014.129.

[67] FEMA Emergency Management Institute. IS-368: Including People With Disabilities & Others With Access & Functional Needs in Disaster Operations. U.S. Department of Homeland Security FEMA; 2022. Available from: https://training.fema.gov/is/courseoverview.aspx?code = is-368\.

[68] U.S. Department of Homeland Security FEMA. Developing and maintaining emergency operations plans comprehensive preparedness guide (CPG) 101 version 3, 2021. Available from: https://www.fema.gov/sites/default/files/documents/fema_cpg-101-v3-developing-maintaining-eops.pdf.

[69] TRAIN Learning Network. About the TRAIN learning network. TRAIN Learning Network (TRAIN); 2021. [cited 2021 Dec 20]. Available from: https://www.train.org/main/welcome.

[70] Pacific ADA Center. Archived inclusive emergency management webinars. Pacific ADA Center; 2021. Available from: https://adapresentations.org/archiveEM.php.

[71] U.S. Department of Health and Human Services Office of the Assistant Secretary for Preparedness and Response. At-risk individuals. Public Health Emergency. U.S. Department of Health and Human Services Office of the Assistant Secretary for Preparedness and Response; 2021. [cited 2021 Dec 21]. Available from: https://www.phe.gov/Preparedness/planning/abc/Pages/atrisk.aspx.

[72] U.S. Department of Health and Human Services. HHS/ASPR. Access and functional needs. TRAIN Learning Network; 2019. [cited 2022 Jan 2]. Available from: https://www.train.org/main/course/1083869/.

[73] United States Department of Health and Human Services Office of the Assistant Secretary for Preparedness and Response. 2017-2022 health care preparedness and response capabilities, 2016. Available from: https://www.phe.gov/preparedness/planning/hpp/reports/documents/2017-2022-healthcare-pr-capablities.pdf.

[74] Centers for Disease Control and Prevention. Public health emergency preparedness and response capabilities, October, 2018. Updated January 2019. Available from: https://www.cdc.gov/cpr/readiness/00_docs/CDC_PreparednesResponseCapabilities_October2018_Final_508.pdf.

[75] Charlton JI. Nothing about us without us disability oppression and empowerment. Berkeley CA: University of California Press; 1998. https://doi.org/10.1525/9780520925441. Available from:.

[76] Kornat M, Micgiel J. The policy of equilibrium and polish bilateralism. Reflections on Polish foreign policy. Jozef Pitsudski Institute: New York, NY; 2007.

[77] Scott J, Coleman M. Reaching the unreached: building resilience through engagement with diverse communities. J Bus Contin Emer Plan 2016;9(4):359–74. 27318290.

[78] Sievers J. Embracing crowdsourcing: a strategy for state and local governments approaching "whole community" emergency planning. State Local Gov Rev 2015;47(1):57–67. https://doi.org/10.1177/0160323X15575184.

[79] Horney J, Nguyen M, Salvesen D, Tomasco O, Berke P. Engaging the public in planning for disaster recovery. Int J Disaster Risk Reduct 2016;17:33–7. https://doi.org/10.1016/j.ijdrr.2016.03.011.

[80] Sobelson RK, Wigington CJ, Harp V, Bronson BB. A whole community approach to emergency management: strategies and best practices of seven community programs. J Emerg Manag 2015;13(4):349. https://doi.org/10.5055/jem.2015.0247.

[81] Schoch-Spana M, Ravi S, Meyer D, Biesiadecki L, Mwaungulu Jr G. High-performing local health departments relate their experiences at community engagement in emergency preparedness. J Public Health Manag Pract 2018;24(4):360–9. https://doi.org/10.1097/PHH.0000000000000685.

[82] Ramsbottom A, O'Brien E, Ciotti L, Takacs J. Enablers and barriers to community engagement in public health emergency preparedness: a literature review. J Community Health 2018;43(2):412–20.

[83] Sammartinova J, Donatello I, Eisenman D, Glik D, Prelip M, Martel A, Stajura M. Local public health departments' satisfaction with community engagement for emergency preparedness. Am J Bioterror Biosecur Biodef 2014;1:1–6.

[84] Donahue AK, Eckel CC, Wilson RK. Ready or not? How citizens and public officials perceive risk and preparedness. Am Rev Public Adm 2013. https://doi.org/10.1177/0275074013506517.

[85] Sperry P. Community participation in disaster planning and the expectation gap: Analysis and recommendations. Richmond, VA: Virginia Commonwealth University; 2013. Available from: https://scholarscompass.vcu.edu/.

[86] Gershon RR, Muska MA, Zhi Q, Kraus LE. Are local offices of emergency management prepared for people with disabilities? Results from the FEMA Region 9 Survey. J Emerg Manag 2021;19(1):7–20. https://doi.org/10.5055/jem.0506. 33735431.

[87] National Association of County and City Health Officials. Follow-up national assessment of the practices, awareness, and health practices inclusion of people with disabilities in local health departments, 2018. Available from: https://www.naccho.org/uploads/downloadable-resources/Health-and-Disability-Followup-Assessment-Report_June-2018_FINAL.Pdf.

[88] Great Plains ADA Center. ADA coordinators. Great Plains ADA Center; 2022. [cited 2022 Jan 8]. Available from: https://www.gpadacenter.org/ada-coordinators.

Elements of individual resilience

5

Jill Morrow-Gorton

University of Massachusetts Chan Medical School, Worcester, MA, United States

Learning objectives

(1) Explain the mental health symptoms and diagnoses found in conjunction with disasters in children, adults and people with disabilities.

(2) Describe the range of persistence of mental health symptoms after disasters and compare the risk factors associated with that persistence.

(3) Explain the factors that are protective against persistent mental health symptoms and disorders post disaster.

Introduction

Studies of resilience examine the ability of individuals and communities to respond to events that occur. Community resilience defines how a group of people that live geographically in proximity to each other and are municipally organized under the same set of laws weather, adapt to and pull through events [1]. Community resilience depends not only on the corporate resilience, but also on the resilience of individuals living in the community. Therefore, understanding the factors that define and promote individual resilience will strengthen the ability to harness these factors in addressing emergency management. First, it is important to define individual resilience.

Defining resilience

In their report titled *An Agenda to Advance Integrative Resilience and Practice: Key Themes from a Resilience Roundtable*, the RAND organization identified resilience as "the capacity of a dynamic system… to anticipate and adapt successfully to challenges" [2]. Applying this definition to individuals, they characterize individual resilience as "the process of, capacity for, or outcome of adapting well in the face of adversity, trauma, tragedy, threats, or significant sources of stress." Fingerle describes psychological resilience in light of an individual's capacity to cope with considerable risks and adversity, resulting in positive adjustment and healthy development [3]. Much of the academic work in individual resilience studies children and their responses to adverse events, including disasters, but also abuse and neglect, poverty and other adverse childhood experiences (ACEs) known to impact physical and behavioral health and functioning even into adulthood [4]. This body of work follows two somewhat divergent pathways. One path looks at the direct response of children to disasters, with some consideration of pre-existing factors, while the other path examines the impact of the pre-existing factors on adult outcomes in terms of work, family, and stability. Each of these paths brings important considerations to the concept of individual resilience and how to use this knowledge to build resilience in children. In addition, there is a small body of work studying adult factors affecting resilience and ways to begin to improve resilience in at-risk adults. This latter work includes some evaluation and guidance about working with populations of people with intellectual disabilities and the similarities and differences in their responses to events. All of this work contributes to the understanding of individual resilience and its integration promotes a broader engagement to help strengthen both individuals and communities.

Linked with the concept of resilience is vulnerability, which the International Federation of the Red Cross and Red Crescent (IFRC) defines in the context of disaster management as "the diminished capacity of an individual or group to anticipate, cope with, resist and recover from the impact of a natural or man-made hazard" [5]. Physical and mental disabilities, social exclusion and isolation, poverty, and lack of political power comprise some of the factors associated with vulnerability, although the mere presence of these factors does not necessarily predict vulnerability. Risk factors may be associated with particular outcomes, but do not definitively determine those outcomes. For example, children with four or more adverse childhood experiences (ACEs) have a sevenfold increase in problematic drug use compared to their peers without such exposure [6]. However, only a small fraction of these children, about 3.5%, will actually have problems with drug use. The remaining 96.5% will not. Thus, ACEs predict an increased risk for certain outcomes, but cannot determine which children will experience those outcomes. Regardless, it is critical to consider these factors and their impact on vulnerability when addressing individual resilience as a risk assessment helps focus intervention before, during and after disasters.

Vulnerability and adverse childhood events (ACEs)

In the initial ACEs work, the Centers for Disease Control and Prevention (CDC) partnered with Kaiser Permanente in 1995 to 1997 to study the childhood experiences and current health status and behaviors of more than 17,000 adults in California [7]. The original study, published in 1998, surveyed seven categories of experiences defined as abuse, neglect and household dysfunction [8]. The experiences surveyed include psychological, physical, and sexual abuse; violence against the mother; or living with household members who abuse substances, have mental illness, suicidal behavior, or were ever imprisoned. This list has been modified over time to incorporate witnessing violence in the home, having a family member attempt or die by suicide, and household instability due to parental separation or divorce or incarceration, and the numbers of ACEs vary depending on the categorization [9]. However, the association of ACEs with poorer physical and behavioral health outcomes later in life has been demonstrated in both the original study and subsequent ones regardless of categorization. The link between ACE exposure and a higher likelihood of negative health and behavioral outcomes later in life, such as heart disease, diabetes and premature death has clearly been made.

The conclusions from the initial ACE study demonstrated that ACE events were common, with more than half of the group having experienced one ACE and 25% of the group having faced two or more [8]. The most common ACEs were substance abuse in the household and sexual abuse. Higher numbers of ACEs consistently correlate with poorer adult health and other outcomes. Four or more ACEs compared to none showed dramatic differences in prevalence of health and behavior risks. When substance abuse, including alcoholism and illicit drug use, was one of the ACEs, it resulted in the largest differences in health and behavior risks with five times more risk in those with four or more ACEs than those with none. However, the proportion of each group with alcohol and substance use in their household was 16.1% in the group with no ACEs versus 28.4% in the group with four or more ACEs. Differences were also found in the prevalence of depression and suicidal behavior related to number of ACEs experienced. More than 50% of people experiencing four or more ACEs also experienced depression, compared to 20% of those with no ACEs. The prevalence of suicidal behavior was 18.3% in the former, but only 1.2% in the latter. Behaviors associated with health risks and their consequences, such as lack of participation in physical activity, obesity, and smoking were also more prevalent in the group with ACEs compared to the group without. The prevalence of the chronic diseases representing many of the leading causes of death and disability in adults and influenced by health behaviors including heart disease, diabetes, lung disease and cancer was also higher in the ACE-exposed group. These chronic diseases both cause disability and disproportionately impact people with disabilities and those with chronic mental health conditions, playing an important role in the context of disaster management.

Much of the work done after the initial ACEs study attempts to better define risks as well as to identify how to mitigate their influence on adult outcomes, including health. ACE exposure represents a subset of factors impacting vulnerability and is not always predictive. Other factors such as poverty, bullying, discrimination, low birth weight, and disability also contribute to vulnerability and poor adult outcomes and may be more predictive [6]. For example, low family income has a stronger relationship to poorer adult health outcomes than any of the ACEs. As well, ACEs are not present in isolation and are more prevalent in populations of people with these other risk factors which complicate identifying the role ACEs play in vulnerability. Nonetheless, ACEs' contribution to vulnerability and role as an important piece of the individual resilience puzzle provides evidence for their consideration in the building of resiliency.

Some advocate for looking further upstream to address the factors that put children at higher risk of experiencing ACEs and using that evidence to push for a broader system approach [6]. This requires addressing many of the social and economic inequities present in communities with the highest level of factors such as poverty and crime and in which children and their families experience more adversity. Many of these factors also contribute to adult disability and poor physical and mental health and increase the vulnerability of that population. These factors should be considered as they may be more predictive for poor outcomes than ACEs. Focusing efforts on prevention of conditions that lead to and coincide with ACEs would not only decrease the exposure of children to these events, but also minimize the health risks that accompany exposure to ACEs. Ultimately, this would improve the health and reduce population vulnerability. Head Start is a program of the US Department of Health and Human Services for low-income children and families that provides early childhood education, basic preventive health services including dental exams, nutrition, and parent education and involvement. Programs such as Head Start have demonstrated significant impacts on child and adult outcomes with regard to education, employment, and improved behavioral outcomes [10,11]. Studies of the impacts of policies that focus on improving early childhood experiences and access to basic necessities such as food and healthcare show that these approaches have a lasting effect into adulthood and improve adult health and economic outcomes [11,12].

Impacts of disaster on children

Research on the impacts of disasters on children generally focus on mental health, the presence of behavioral symptoms and their dissipation over time. Measures of symptoms of anxiety, depression, oppositional behaviors and post-traumatic stress symptoms (PTSS) such as intrusive memories and flashbacks, avoidance of places and people associated with the event, negative alterations in cognition and mood, and hyperarousal reflect the magnitude of response to the disaster [13,14]. The development of mental health symptoms correlates with various factors associated with the disaster itself such as proximity to the event, injury and other consequences of the event [15–17]. Other factors contributing to mental health and behavior such as prior trauma or stress are also considered in many of the studies. In addition, the role of the family's influence on the child and the impact of the child-family (usually parent) interactions on behavior post event is considered [18,19].

Factors associated with exposure to the event

Proximity to the event has been shown to be proportionally associated with more severe and long lived symptoms of post-traumatic stress in children [17]. Event exposure is defined in a number of ways, but it generally refers to the physical closeness of the child to the event and the physical impact of the event on the child. Proctor and colleagues assessed young children who experienced an earthquake defined exposure by its effect on the children including physical injury, damage to their home, displacement from home or school, loss of possessions, and financial difficulties and stressors [20]. Hardin described that in addition to the degree of actual threat to the child from proximity to the disaster and physical injury, experiences witnessing the event also contribute to the risk of developing Posttraumatic Stress Disorder (PTSD) with the persistence of Posttraumatic Stress (PTSS) symptoms over time [21]. Others have shown that children's subjective responses to the event can also predict later outcomes [22–24].

Children often exhibit their response to disasters through behavior. This response can be expressed in varying ways, but there are some themes characterizing those responses. Symptoms of PTSS are commonly seen with close proximity to the event. In the 2 to 6 weeks after the 1989 Loma Prieta earthquake, about 50% of 2 to 3 years old children displayed clinginess and separation anxiety related to their parents [15]. Young children often seek proximity to their parents when stressed and they demonstrate increases in this behavior following disasters. Studies of hurricane impact on the behavior of preschool and early school aged children indicated that they expressed their distress by repeatedly incorporating the event into their play and conversation. Children have also have become aggressive, depressed, withdrawn and have regressed developmentally when separated from their parents during an emergency, as was observed in London during the Blitz in World War II. Both separation from parents and alteration of their relationship with their parents can impact children's behavior. After the 9/11 terrorist attacks, some preschool children whose mothers developed depression and PTSD demonstrated increased aggressive behaviors and emotional reactivity [25]. As well, sometimes the behavior expressed is physical in nature. Following a Bangladeshi flood, 34% of children 2 to 9 years old showed regressive behavior in the form of enuresis [15]. Thus, the range of behaviors seen in response to trauma and disasters in children are varied both in expression and age at which behaviors are expressed.

Conceptual models of children's responses to disasters generally contain several potential factors related to the characteristics of exposure to the event, child characteristics prior to the event such as trauma experience, and post-event factors [15–17]. In particular, the models consider the impact on young children for whom early childhood represents a period of active development as well as a period during which concepts such as death may have little or no meaning [15,26]. Child development is a continual process over time, progressing through childhood into adolescence and beyond, where children develop the capacity to function and think as adults over time. Deering applied the cognitive developmental theory to how preschool, school-aged children and adolescents respond to and process disasters with a focus on better understanding age-specific responses [27]. Recognizing that these differing responses are based on developmental level is important in appropriately determining effective intervention to address both immediate and post disaster responses. However, results of studies of the impact of age and gender on post disaster outcomes on mental health have been mixed. Hafstad and others have shown that children's age does not necessarily predict outcomes in a disaster and children of any age can experience an adverse outcome post disaster [18,28–30]. Likewise, results of these studies that examined the impact of the child's gender on outcome did not show a definitive effect. On the other hand, King studied the impact of multiple disasters on children over age eight in Louisiana with regard to age and gender, and identified that both younger age and female gender were associated with additional variance that predicted the presence of more PTSS symptoms [16]. Therefore, gender and age may not be as important in predicting post disaster outcomes as other factors.

Unlike age and gender, previous trauma experience has clearly been shown to significantly increase the risk of poor outcomes post disaster [31]. This trauma fits into two different types: exposure to previous disasters and intrapersonal adverse events or trauma such as ACEs, poverty, and others. Both have been studied individually and together to assess the role of trauma and the magnitude of the impact of trauma on outcomes for children. These factors have been examined with an eye toward protective and adverse characteristics in children's lives that might help mitigate the trauma and build resilience. They have been studied in light of a target disaster both proximally in time to the disaster and over time to get a sense of the symptom persistence and longer term mental health impact. As well, the longer term studies have evaluated the impact of adult responses to the disaster on children's outcomes.

Predictors of mental health outcomes

Cloitre and colleagues concluded that PTSS and PTSD can be predicted by childhood but not adult cumulative exposure to trauma [31]. While age may not consistently predict specific responses to a disaster, cumulative exposure to trauma in the first years of life may have adverse biological effects on the developing nervous system, impacting development and increasing mental health risks [18,32,33]. Mullet-Hume and colleagues investigated lifetime trauma in middle school students exposed to the 9/11 attack on the World Trade Center [34]. They found that a history of cumulative trauma exposure in everyday life posed higher risk for mental health problems than severity of exposure to the disaster. Studies of outcomes such as that by Catani and others identified that prior exposure to events related to war and family violence contributed to poorer adaptation post disaster following the 2004 Indian Ocean tsunami [28]. King identified that interpersonal trauma experienced by children, including events such as neglect, abuse, domestic and community violence, and victimization, significantly predicted additional elevation in PTSS within a subgroup that had higher baseline PTSS symptoms [16]. In a similar manner, Kronenberg and colleagues demonstrated that presence of the experience of additional trauma by disaster-exposed children changes the trajectory of risk for longer term sequelae and helps explain the variable recovery from a disaster [29].

The literature shows that children and youth express different patterns of recovery after being exposed to a disaster. Kronenberg's longitudinal follow-up of children and adolescents ages 8 to 18 exposed to Hurricane Katrina identified four patterns of recovery related posttraumatic stress symptoms and disorder [29]. The largest subgroup, approximately 45% of the study sample, called the "stress resistant" group showed no significant response to the disaster in terms of PTSS symptoms. In the next most prevalent response group, referred to as the "normal response and recovery" group, 27% of the children and youth had initial symptoms that resolved over time. The "delayed breakdown" group, representing 5% of the sample, showed a delay in the appearance of symptoms. The fourth group, making up almost a quarter of the children and youth (23%), falls into the "breakdown without recovery" category. Individuals in this last group developed PTSD symptoms that did not resolve over time. Similar patterns of recovery have been conceptualized in adults, although the patterns have different names. Bonanno and Mancini hypothesize four adult post-disaster recovery patterns: (1) resilience with an absence of elevated PTSD symptoms; (2) recovery with initially elevated PTSD symptoms that return to adaptive functioning levels: (3) chronic dysfunction where elevated PTSD symptoms do not abate with

Table 1 Labels for patterns of recovery post disaster in adults and children and youth [33,35–38].

Patterns Kronenberg [29]	Patterns Bonanno and Mancini [33]	Patterns Other authors [35–38]
Stress resistant	Resilient	Resilient
Normal response and recovery group	Recovery	Recovery
Delayed breakdown	Delayed trauma	Persistent stable or worsening symptoms
Breakdown without recovery	Chronic dysfunction	

time; and (4) delayed trauma characterized by an increase in PTSD symptoms over time with elevated levels [33]. The comparison of these two patterns appears in Table 1.

Other researchers have also identified differential responses to recovery post-disaster exposure. Self-Brown, Bokszczanin, Goenjian, and Baddam each worked with their teams on an activity similar to Kronenberg's work to categorize the various patterns of response to disasters into groups [29,35–38]. Each of these research teams described three, rather than four, categories of disaster response patterns, the first two of which match the naming convention posed by Bonanno and Mancini [33]. These include a resilient group that doesn't develop symptoms, a recovering group that develops symptoms that resolve over time and a persistent group with worsening or persistent symptoms. The latter appears to combine the delayed breakdown and breakdown without recovery groups described by Kronenberg and the chronic and delayed trauma groups into a single group with persistent stable or worsening PTSS symptoms over time as shown in Table 1. These patterns of differences in responses not only transcend disaster types, but also reflect the presence or absence of pre-disaster stressors with the recovery patterns of youth who had experienced trauma and loss before or after the hurricane more likely characterized by persistent and worsening symptoms [29].

Another factor impacting response to disasters relates to prior disaster experience. While some disasters represent one time events, many areas are disaster-prone and frequently experience repeated natural disasters such as earthquakes, hurricanes, and cyclones. The World Atlas compiled a list of the 10 most disaster prone states in the United States based on the number of disaster declarations [39]. Topping the list is Texas followed by California, Oklahoma, New York and Florida. Reflected in the top 10 are 4 states with coast along the Gulf of Mexico: Texas, Louisiana, Florida, and Mississippi. Of these, the impact of experiencing multiple disasters has been most studied in Louisiana, particularly related to the aftermath of Hurricane Katrina. Within a 5 year period, Louisiana experienced two of the worst disasters in US history: Hurricane Katrina in 2005 and the Deepwater Horizon Oil Spill in 2010. Both of these events resulted in widespread destruction, property loss, and disruption to family life leading to short and long-term symptoms of PTSS [15,16,31]. The cumulative impact of these two events was associated with increased negative responses consistent with a dose-response effect where increasing levels of exposure or dose result in more negative responses [40,41]. Weems and colleagues have demonstrated that the effect of disaster-related stress and trauma on youth with prior disaster exposure has disproportionate consequences and those with the highest exposure to both exhibit the highest PTSS scores [42,43]. Contributing to this are studies putting forward the hypothesis that for the oil spill, in particular, the chronic stress related to the uncertainty of financial well-being and health and environmental risks, may be more detrimental to mental health than the direct impact of the actual event [44].

King and colleagues studied children and youth from Louisiana who experienced both Hurricane Katrina and the Gulf oil spill to identify the factors impacting their mental health in the short and long-term after these events [16]. Through questionnaires and interviews the authors determined the presence of PTSS symptoms in light of the experienced disaster-related and intrapersonal trauma including many ACEs events. While about a quarter of the children had experienced intrapersonal trauma, almost all of them, 89%, endorsed experiences related to Hurricane Katrina and 79% reported consequences of the oil spill. Higher PTSS symptoms occurred in children with hurricane exposure, oil spill stress, younger age at exposure, female gender, and minority status which explained about 16% of the variance in symptoms. An additional 4% of the variance was associated with those factors plus an added element of intrapersonal trauma or loss. Timing of the experienced trauma did not always precede the disaster as about 15% experienced trauma prior to both events, 40% post Katrina and 45% after the oil spill, reflecting the often on-going nature of intrapersonal trauma. This work illustrates that intrapersonal trauma and loss interwoven with multiple exposures to disasters over time results in mental health symptoms, including PTSS, that may persist into adulthood. Children and youth living in communities subject to disasters are vulnerable to the effects of the ongoing stress of recurrent disasters coupled with individual trauma and loss. These elements impact the resilience of individuals, but also identify some factors that offer the opportunity to create resilience.

Effect of disasters on children in their families

Studies of mental health outcomes in children and youth from exposure to trauma, including disasters, also examines the impact of protective factors. Children experience trauma and loss in the context of their families as well as their communities. Often children and their families are directly exposed to the same disaster events, although direct exposure is not required to experience trauma. Children may be indirectly exposed to an event experienced by a family member and behavior of family members in response to this event may impact children negatively [43]. Therefore, children's outcomes particularly as they relate to mental health and PTSD depend not only on their personal experiences of disasters and interpersonal trauma, but also on the experiences and responses of their family to those same events and events in their past.

In the 1970s, Bronfenbrenner developed the Ecological Systems Theory to look at child development from a broader view in response to studies of child behavior and development examining the impact of a parent on the child in a laboratory setting [45]. He noted that children are part of a larger environment and those factors have an impact on the development of the individual child and the child has an effect on the people in their environment. The theory proposes five nested, interdependent ecosystems with the child in the middle and the progressively larger and less proximal systems impacting the child. This theory has been used to explain interactions related to disaster response [46–48]. These include the microsystem, mesosystem, exosystem, macrosystem and chronosystem, and the relationships between each are reciprocal. Fig. 1 illustrates a simplified version of the concept and the interactions between the factors in each ecosystem. The ecological model has been used to better understand child development and behavior in the context of environment, but also has been applied at the community level to better understand behavior within a community and to improve individual and collective resilience [49][48]. Thus, using social ecological and family systems theory better helps explain the

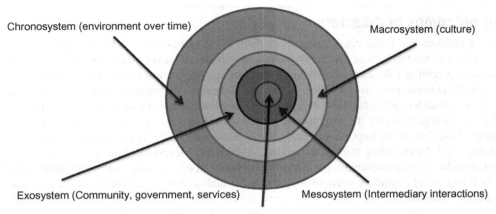

Chronosystem (environment over time)

Macrosystem (culture)

Exosystem (Community, government, services)

Mesosystem (Intermediary interactions)

Microsystem (child and immediate contacts)

FIG. 1

Simplified Bronfenbrenner's ecological systems theory.

Adapted Kerns CE, Elkins RM, Carpenter AL, Chou T, Green JG, Comer JS. Caregiver distress, shared traumatic exposure, and child adjustment among area youth following the 2013 Boston Marathon bombing. J Affect Disord 2014;167:50–5.

response to and impact of disasters on children in the context of their surroundings including parents, families, and community.

This principle applies to adults as well and there is a large body of evidence showing that symptoms in disaster survivors also affect others around them. The post-disaster adjustment period experienced by parents and children is experienced as an interdependent unit, or dyad [18,19]. This is especially true of dyads with young children because physical proximity for performance of daily care such as dressing and feeding is more prominent with young children. Juth studied PTSS and distress levels in parents and children after an earthquake in rural Indonesia in order to determine the impact of each on the other [19]. This research showed that PTSS symptoms best predicted general distress in both children and parents. In terms of the interactions between children and parents, parents whose children experienced higher distress had more PTSS symptoms compared to those whose children had less distress. The converse did not hold true and parental distress did not predict symptoms of PTSS in children. Other researchers, however, have found that children of parents with high levels of distress are at increased risk for psychopathology and that in families with already existing family conflict and domestic violence the increased trauma of a disaster compounded the family stress [48]. Scaramella and colleagues established that the impact of financial hardship and other hurricane-related stressors on the family, particularly on mothers, impacted the psychological outcomes of young children [49]. Increased family stress was associated with maternal depression leading to less effective parenting and behavior issues in the children. Studies of families with children and the impacts of the Boston marathon bombing in 2013 demonstrated a similar outcome in older children as well [50]. Increased parental distress resulted in increased posttraumatic stress in older children causing emotional problems, hyperactivity and inattention, difficulties with peers, and lower levels of prosocial behaviors such as helping, sharing, and cooperation. In general, high levels of caregiver distress have been associated with the development of PTSS and behavioral issues in children post exposure.

Adult outcomes of disasters

Studies of children and the effect of disasters on them in the context of their families give a small window into the potential impact of disasters on adults as parents. However, adults also respond to disasters outside of a parenting role, and understanding that fact can lead to ways to prevent and mitigate those responses. In addition, prior to examining the impact of disasters on populations of people with disabilities it is important to understand how their responses are both similar to and different from the general population. Factors affecting the response of people with disabilities to disasters are both unique and common. These factors are important to recognize as these may be the key to building resilience in this population and fully including them in the process. Aldrich and Meyers maintain that social capital or the social networks within communities are critical to resilience in the response to and aftermath of disasters [51]. Building social capital by strengthening the individuals that make up the collective or community requires understanding the underlying factors contributing to and impeding resilience and strength.

Fewer studies examine the effect that disaster has on adults in the detail contained in the child studies of events such as Hurricane Katrina and the Boston marathon bombing. Fernandez pulled together literature investigating one of the most common disasters associated with multiple causes, flooding, and mental health outcomes in adults [52]. Multiple studies show that the presence of flooding is associated with a number of different mental health outcomes. Adults experiencing floods have increased PTSS symptoms, anxiety and depression, compared to those without such experience. PTSS symptoms are postulated to be related to development of other mental health conditions. Both children and adults, with the exception of those older than age 70, had an increased incidence of developing depression associated with flooding events. Results related to other conditions such as suicidality and substance use did not show a consistent association as they were increased in some studies, but not in others. Mental health issues stemming from floods occur in both the developed world and developing countries and are associated with low socioeconomic status and material deprivation.

Boscarino and team noted that disasters caused by large scale events with characteristics of loss of life and significant economic disruption such as hurricanes can result in significant psychiatric disorders [53]. They studied adults from New Jersey who lived through Hurricane Sandy in 2012 in order to determine the impact of the hurricane on mental health and the use of behavioral health and other services. Hurricane Sandy was the worst disaster in New Jersey, with property destruction similar in magnitude to Hurricane Katrina in New Orleans. The adults studied lived in the hardest hit small- and medium-sized beach communities. The study population had high rates of disability and lifetime stressors characterized by 17.5% of the population having physical health limitations, 15% having hurricane related experience and 15% having high exposure to lifetime stressors and traumatic events. About a quarter of the sample accessed hurricane-related financial assistance and almost a quarter self-described as having low social support. Six months after the hurricane, 14.5% of the sample had a positive PTSD screen and 6% met diagnostic criteria for major depression. Predictors for mental health outcomes included high exposure to stressful life events, low social support, high storm exposure, and receiving financial assistance. Additionally, physical health limitations were associated with depression, but not PTSD. In this population, as in the flooding studies, older age was generally protective related to developing mental illness [52,53]. About 20% of the population sought mental health support and 14.5% participated in counseling. These results were similar to findings after the New York 9/11 terrorist attacks in the short run, although delayed PTSD and the seeking of mental health treatment after 9/11 increased substantially by 2 years post event [53].

Longer term studies of adult outcomes

The literature generally focuses on the short term post event outcomes, although some longer term studies exist. Paxton and colleagues studied PTSS symptoms and psychological distress in a sample of New Orleans mothers with low incomes from three points of time: 1 year prior to Hurricane Katrina, and 7 to 19 months and 43 to 54 months after the event [54]. In this vulnerable group of women, both PTSS and psychological distress decreased over the two post-Katrina study periods, but remained high, never returning to normal. In addition, similar to children, the responses of these women paralleled the proximity of the exposure to the disaster and those with more hurricane-related traumas had a higher risk of developing delayed PTSS and chronic mental health problems regardless of their psychological distress level. Factors predicting the development of PTSS with psychological distress included psychological distress before the hurricane, hurricane-related home damage, and exposure to traumatic events. The researchers found that home damage was more highly predictive of PTSS and hurricane-related trauma and pre-hurricane poor mental health or low socio-economic status was more highly associated with post-event mental health complications. In this study, higher income and greater social support protected against the development or worsening of mental health problems.

Studies of other disasters have shown similar findings to those related to Katrina. Beinecke and others demonstrated through interviews of people who experienced the Boston marathon bombing that while most people recovered from the event, about 10% to 15% developed severe mental health outcomes including PTSD [55]. This group identified other psychological outcomes in both survivors and first responders. They found symptoms of distress responses with insomnia, irritability, a sense of vulnerability, distractibility, mental health issues including PTSD, depression, and complex grief and behavior changes involving smoking and alcohol use. First responders also exhibited over-dedication to their work. More recent studies of the Boston Marathon bombings show that not only do the experience of both childhood and adult traumas contribute to poorer functioning in adults, but also these negative life events may influence the collective response to an event such as terrorism [56]. Hobfoll who also studied survivors of terrorism notes that cognitive and emotional issues such as anxiety, depression, irritability, and loss of intellectual capacity can follow traumatic experiences [57]. Other behaviors may ensue including sleep problems, social withdrawal or substance abuse [58]. Physiologic signs of psychosomatic reactions, such as increased resting heart rate, were also found and that could imply a negative impact on physical health as well. Risks of mental health and trauma include life experiences like abuse and loneliness. Negative mental health outcomes in adults are not only associated with disasters, but complicated by life experiences such as living in areas with high rates of poverty and violence, abuse, loneliness and influences responses to disasters.

One of the longest post disaster studies, conducted by van der Veldan in the Netherlands, examined mental health outcomes in people who experienced a major explosion and fire in a fireworks factory in a residential area in the city of Enschede, the Netherlands [59]. This fire resulted in significant impacts to the neighborhood with damage to about 500 homes, 23 deaths and more than 900 injured people. The van der Veldan study looked at the development of persistent mental health disturbances using the following measures: anxiety and depression symptoms; sleeping problems; and the use of physician-prescribed tranquilizers. Evaluations occurred at four periods of time longitudinally from 2 to 3 weeks to 10 years post disaster. Outcomes showed that mental health symptoms persisted in approximately 10% of people who displayed severe symptoms 2–3 weeks after the event. In the long term, affected residents had more chronic anxiety symptoms and sleeping problems than controls and those with high levels of exposure to the disaster were at higher risk of persistent PTSD, anxiety, depression and sleep problems. Physician prescribed use of tranquilizers did not differ among the groups.

Patterns of development of mental health symptoms and disorders

Other studies of recovery of adults after a disaster examine the patterns of the development and resolution of mental health disorders and symptoms. Typically, the majority of people exposed to disasters do not experience elevated mental health symptoms [60]. For those who develop these symptoms as Kronenberg describes in children, researchers have identified a few different trajectories [29,60,38,61]. Self-Brown and colleagues defined three trajectories including resilient, recovery and chronic groups in mothers exposed to Hurricane Katrina [38]. The majority of mothers, 66%, fit into the resilient group, with 30% and 4% in the recovery and chronic groups, respectively. Two years after Katrina the proportion of mothers in each group was roughly unchanged, although the chronic group had increased to 10% possibly indicating some level of development of late symptoms. Most mothers had few symptoms at that point in time. The mothers studied reported exposure to 2.15 traumas on average. Prior exposure to traumatic events was predictive of being in the recovery group rather than the resilience one and social support was protective for mothers.

Lowe and colleagues also studied the trajectory of recovery of mothers with low incomes over a period of 4 years post Katrina and identified six distinct groups [60]. Using Kessler's K6 Scale, a six-item screening measure of non-specific psychological distress, the researchers studied three time periods post event, 1 year, 2 years and 4 years, and applied latent class growth analysis, a methodology to identify subpopulations within a heterogeneous population, to define the six groups [60,62]. The following characterized the groups: (1) resilient with low scores across all time periods; (2) coping with moderate, but not abnormal, scores across all three time periods; (3) increased distress with high scores across all time periods; (4) delayed distress with low scores in the first two time periods and an increase in the third; (5) decreased distress with high initial scores and then dropping, but not to normal; and (6) improving with high initial scores that drop to normal. The majority of the mothers fit into the resilient and coping groups with 62% and 22%, respectively. The proportion of mothers in the other four groups was similar, ranging from 3.2% in the improving group, 4.5% in the delayed distress group and 3.9% in each of the other two.

The authors identified some key risk factors for poorer outcomes including childhood trauma, more recent traumas and stressors associated with higher initial distress scores, and intimate partner relationship issues such as drug and alcohol use, incarceration, and conflict all contributed to both pre- and post-disaster distress. This study, like the previous ones, identified that recovery from disasters and the risk of developing long term mental health symptoms is defined by factors present prior to the disaster as well as disaster and post-disaster experiences.

Experience of disaster in people with disabilities

As a subgroup of the general population, people with disabilities experience disasters with the rest of the population. Some of the adult studies examined the effect of poor mental health or physical health prior to disaster on recovery in the general population. However, as studies of adult outcomes of disasters are limited, studies of outcomes in people with disabilities are even more limited. Jonkman and

Kelman noted that the majority of injuries and deaths occur in people with lower incomes, chronic diseases, racial and ethnic minorities, or limited English proficiency and those with age extremes such as the elderly or very young [63]. Many people with disabilities live in poverty or relative poverty, have multiple chronic medical and behavioral health conditions, have limited access to transportation because of physical limitations, and lack access to health and education resources. Because of their low incomes and limited financial resources, they often live in housing and neighborhoods at increased risk for damage or destruction during a disaster. The availability of home and community services encourages people with disabilities to live in the community outside of institutions, enabling them to be an active part of their families and communities. They are dispersed throughout the community, some of them living alone and others with family. Many require mobility assistance and consequently need help to evacuate. However, being spread out across a region or community makes getting evacuation assistance to them more difficult. While some of these factors increase vulnerability in people with disabilities in the short run, awareness and understanding of the factors can promote building resilience in the longer term.

Disaster studies that have assessed disability status generally identify that roughly 15% of the people surveyed had been told that they have a disability and somewhere between 15% and 25% were unable to evacuate either because they had a disability or were caring for someone with a disability [64–66]. In a sample of more than 500 people over 65 years old living in southern Louisiana, McGuire and colleagues found that 31% had a disability with 16.6% of them using mobility or other assistive devices such as a cane, wheelchair, hospital bed or modified telephone [67]. Physical health and mental health outcomes differ in populations of people with disabilities. Studies have shown that older individuals have a higher risk of physical injury from a disaster. Younger people tend to have more psychological or mental health outcomes than older people, which is consistent with studies of adults where older people appear to be more psychologically resilient [53,54,38]. Others at higher risk include women who are the heads of their household, people with low socioeconomic status, and members of racial and ethnic minorities [38].

Rooney and White queried 56 people with disabilities from 20 states about their experiences with disasters and both the issues and helpful activities that occurred [68]. Eight of the participants (13%) noted some experiences of fear, grief, nightmares, and generalized stress after the disaster. The group identified six issues that they experienced related to disaster response. In general, there was a lack of workplace and community planning and people with disabilities were often left behind while people without disabilities were evacuated. Shelters and temporary housing options lacked accessibility and were unable to meet the needs of people with disabilities. As well, the disaster relief personnel were unfamiliar with the potential issues that people with disabilities face and unaware of disaster relief for them. Infrastructure, such as public transportation, potable water supplies, access to elevators and power supplies did not meet the needs of people with disabilities and were unavailable to them. The last issue noted was difficulty in returning to everyday activities. These factors, while experienced to some extent by all disaster survivors, significantly impact people with disabilities, leaving them without basic functional abilities. Lack of power interferes with the ability to charge a power wheelchair, which for someone who cannot walk results in the inability to get from place to place. People who depend on power for devices such as nebulizers to take their respiratory medication may have a preventable worsening of their medical condition that could result in hospitalization or death if access to medical care is not available. As well, many people with disabilities do not drive and are reliant on public transportation. However, public transportation often becomes non-functional or functions in a limited capacity during and after disasters, impacting their ability to evacuate or to access relief services and supplies.

Post-disaster experience of people with intellectual disabilities

Individuals with intellectual disabilities (ID) face similar issues in the aftermath of a disaster. This population has varied strengths and needs because of differences in the range of cognitive functional levels impacting their response to disasters. However, people with ID experience disaster-related trauma and grief analogous to others [69]. This is often coupled with a history of trauma from both childhood and adulthood, complicating the disaster response and outcome as it does in others with trauma exposure. The experience of trauma in people with ID on average probably exceeds that found in the general population. Valenti-Hein and Schwartz found that a little less than half of the people with ID in their sample experienced more than 10 episodes of abusive trauma in their life [70]. These combined with the losses and trauma that occur with a disaster increase the potential vulnerability of this population and the potential for a poorer outcome post disaster. In addition, people with ID often have cumulative losses throughout their life. Those who live in communal living arrangements lose privacy and the relationships of living with family members. As well, for some, family members may no longer participate actively in their life, creating additional loss. Health problems related to the cause of their ID or acquired conditions like high blood pressure and diabetes also create functional impairments and limitations. Some people with ID require a structured environment and have difficulty coping with even minor changes in the environment. Disruptions related to moving into a new home cause these individuals significant distress. All of these factors contribute to the experience of people with ID in disasters, but do not preclude resilience and the ability to adapt to and cope with these events.

People with ID experience trauma and loss, but may exhibit it differently. Takahashi and colleagues described the experience of people with ID in Japan after the Hanshin earthquake [71]. The segregation and exclusion of people with ID led to dehumanization and feelings of powerlessness, furthering vulnerability and hindering resilience and recovery. Like others, people with ID experience depression, PTSS, and complicated grief in response to a disaster. They feel anger, confusion, fear and anxiety during and in the aftermath of a disaster. However, they may manifest this response in behavior that differs from what is typically seen in people without ID. Self-injurious behaviors and somatic complaints are common as are increases in compulsive behaviors, frustration, hyperactivity and irritability. Symptoms of PTSD may be expressed as agitation, physical aggression, or disorganization in addition to more typical behaviors like sleep difficulties or depressed mood. The intensity of the behavior and the disruption it can cause often lead to misinterpretation of its etiology, mistaking the behavior for mental illness and not a reaction to an event. As well, these behaviors are viewed as problematic, but frequently not recognized as expressions of trauma and loss [71]. Despite these differences in expression of distress, people with ID benefit from the same range of interventions as the general population, albeit with some modifications in delivery. Resilience can be fostered in this population to reduce the potential impact of a disaster on mental and physical health outcomes.

Resilience in individuals, children and families

Resilience results from the interplay between negative and protective factors on psychological functioning. Both the magnitude and proximity of trauma contribute to the effect on resilience as does the experience of childhood trauma. Life stressors such as poverty, social isolation, and poor mental and physical health contribute negatively to the recovery ability of people exposed to disasters. The added burden of multiple traumatic events further complicates an individual's ability to be resilience in the face of a disaster. Despite all of these negative factors, protective factors exist that can lead to resilience with less psychological distress and fewer mental health symptoms. Studies show that while a number of people are negatively affected by disasters, most children and adults demonstrate resiliency without persistence of symptoms of mental health disorders [60]. In Kronenberg's longitudinal follow-up of children and adolescents aged 8 to 18 years exposed to Hurricane Katrina, the "stress resistant" group which displayed no significant ongoing response to the disruptions of the hurricane was the largest subgroup representing almost half of the group [29]. This subgroup combined with the "normal response and recovery" subgroup indicate that over 70% of children and adolescents responded to Katrina without developing evidence of long term mental illness symptoms. Identifying the factors that protect individuals from adverse outcomes and allow them to adapt and cope may be a start to determining how to build resilience in individuals and communities before a disaster strikes.

Bronfenbrenner's Ecological Systems Theory defines impacts on children's behavior in the context of their parents, family and community [45]. Children's responses to disasters often reflect their parents' level of stress and parents in turn mirror their childrens'. A broader view of the Ecological Systems Theory applies to adults in disaster settings as well. Studies show that resilience is linked to social connections and relationships both pre- and post-disaster [46,48,49]. The relative impact of relationships reflects the proximity to the person, with intimate and family relationships having a greater influence than community ones. However, community and social connectedness are important factors in individual and community resilience. Studies of child and adult disaster experiences reveal some of these factors.

Hafsted studied Norwegian parents and children after the 2004 tsunami and noted associations between the number and severity of PTSS symptoms and proximity to the event experienced by the parent-child dyads [18]. Findings of a strong positive association between parental and child PTSD have been consistent in the literature, however, children with a strong, supportive family environment adjust better than those without [15,18,48]. Parent support and positive family functioning have been found to buffer children's reactions to disasters and trauma. Juth and Hafsted studied dyads of parents and children that experienced tsunamis in rural Indonesia in 2006 and Southeast Asia in 2004, respectively [18,19]. Juth examined poor rural Indonesian families that remained in place during and after the disasters, while Hafsted analyzed behavior in relatively affluent Norwegian families who were evacuated back to Norway post disaster. Both studies outline similar conclusions with regard to the behaviors and responses that led to the better outcomes for the children. In particular, the evidence suggests that there is an important role in the unidirectional effect of parents' responses on children's mental health post disaster. Hafsted expands this to identify a framework of the effective, positive ways that parents supported their children after the event [18].

Positive parent behaviors reflected an ability to modify support to meet the child's individual needs [18]. Successful parenting strategies included returning to a normal routine as soon as possible and establishing a sense of safety and emotional support for the child. In addition, parents implemented a strategy termed "scaffolding" where the parent builds a metaphorical scaffold or support structure around the child allowing them to use their own coping strategies and intervenes to provide support only when those strategies are ineffective. This approach reflects the guidelines for how parents should help their children after disasters developed by the National Child Traumatic Stress Network (NCTSN) and National Center for PTSD (NCPTSD) [18]. These guidelines promote focus in three areas: a sense of safety, a sense of self-efficacy and connectedness, and hope. One of the strategies used to achieve this is to invoke calming procedures. Ofsofsky and colleagues identified that children's relationships with their caregivers diminished the impacts of disaster-related post-traumatic stress [15]. In particular, both positive and negative caregiver behavior during play was significant for all children and adolescents, not just young ones [15,29]. Thus, a relationship with a supportive, caring adult to assist in navigating the experience of a disaster can help buffer children from the potential negative effects of a trauma.

Integral to the role of a parent in supporting their children through a disaster is managing their own response. The function of the dyad depends on the mental health and distress levels of the parents and Kerns found an association between highly distressed caregivers and child post-traumatic stress symptoms after the Boston Marathon bombing [50]. Research suggests that the family unit provides a potential structure to use to focus interventions for both children and adults [48]. Factors and interventions that promote resiliency in adults can minimize parents' distress and help them be more available to their children. More resilient adults also contribute to better community and collective resilience and better outcomes post disaster. Studies of adults show that, like children, most adults are resilient in the face of a disaster, but that there are some predictive factors for resilience [71,75]. Bonnano studied a random sample of New Yorkers after the 9/11 disaster and identified a number of factors relevant to resilience [72]. In this study, resilience was defined as absent or minimal PTSS symptoms, depression and substance use. In addition to factors already identified to predict the presence of negative predictors, including higher level of trauma exposure, female gender, and recent and past experiences of disaster or trauma, the researchers noted education level, income change, social support, and occurrence of chronic disease as predictors. Some of these factors offer potential for mitigation and intervention both before a disaster as well as afterwards. Additionally, Seery and others postulate that the development of resilience may represent a U-shaped curve where both those individuals with little experience of trauma or adversity and those who experience a great deal are less able to cope with the events than those with some experience with negative life events [73,74]. Thus, people who experience some level of adversity have better outcomes than those with too much or none and some level of adversity may teach skills that help these individuals deal more effectively with the consequences of a disaster.

Most research on pre- and post-disaster factors that predict outcomes focus on those with negative impacts and poor outcomes. However, in order to better understand resilience and be able to build it, it is important to know the protective elements as well, as these may help prioritize the areas for development of effective interventions. Höfler, in a review of psychological resilience observes that adults have the potential to build resilience and coping strategies to use in the event of a disaster [58,61]. This resilience study examined managing disasters from two different perspectives which assume outcomes of maintaining positive health and returning to positive psychological functioning in a short period of time. Factors fall into two different areas: personal factors related to the individual and environmental ones outside of the person. Addressing each of these areas has the potential to improve not only individual resilience, but also community conditions affecting resilience in the population.

Researchers have identified factors associated with positive adaptive responses to disasters and trauma. Höfler notes that optimism, having a purpose in life, internal control, and self-efficacy can reduce dysfunctional responses to trauma [58]. Multiple researchers have shown family and institutions to be protective as are social supports including positive role models [58,60,75]. In flooding studies people's sense of identity was found to be integrally connected to people's house and neighborhood, underlining the importance of quickly reintegrating people back into their community [52]. Religion and involvement in a religious community proved a protective factor for a number of reasons [60]. In addition to providing opportunities for prayer and attending services, religious organizations were sources of needed physical support including food, shelter, and financial help during and after disasters. Social connections to and relationships with other members of the congregation also support people after a disaster and protect against maladaptive responses. Other researchers who studied the aftereffects of 9/11 in New York City and Hurricane Katrina in New Orleans similarly recognized the protective role of social connections within a family, neighborhood or community in maintaining resilience [41,60]. In another study post 9/11, Sherman and colleagues identified an additional factor related to preparedness that lead to better psychological outcomes post disaster [76]. Higher emergency preparedness safety climate scores on the European Process Safety Centre (EPSC) disaster preparedness measurement tool were associated with better outcomes and resilience, making this tool potentially useful in measuring factors that contribute to resilience. Factors that are either modifiable or help match groups to target interventions provides opportunities to most efficiently use limited preparedness funding to achieve the best outcomes.

Conclusion

Risks and protective factors shape not only acute symptoms, but also longer term mental and physical health. Populations at particular risk for both short term and persistent symptoms include those with chronic physical health conditions, disabilities, pre-existing mental health conditions, and economic disadvantage. While even in high risk populations not all acute symptoms lead to poor outcomes, many people require some level of intervention to help effectively resume day to day functioning post disaster. Evidence shows the benefits of helping people get back to their everyday lives and protective factors can mitigate the impact of disaster on mental and physical health. Building individual resilience can moderate some of the factors leading to poorer outcomes as well as offer some coping strategies that can be used throughout life, not just during disasters. Improving individual resilience can contribute to fostering community resilience and in turn, stronger communities better able to support their members in difficult times. Interventions to decrease the vulnerability of populations, including addressing modifiable factors, targeting higher risk populations such as people with disabilities and women, and incorporating both personal and environmental aspects of vulnerability into an intervention can help diminish acute symptoms and shield populations from long term mental health conditions and the consequent physical health effects of disasters. Counteracting this vulnerability and building resilience requires: (1) using planning and preparedness tools to reduce the effect of the disaster; (2) giving people coping strategies to manage during and after a disaster; and (3) confronting the root causes of risk and vulnerability, including poor access to resources, social isolation, poverty, and adverse childhood experiences. Capitalizing on factors that lead to vulnerability and those that create resilience will help people with disabilities improve adaptive responses to both daily life and disasters.

Questions for thought

(1) Consider, as an emergency planner, how to support the evacuation needs for the population of people with mobility disabilities living in the community. Identify what actions would support this population with regard to evacuation.

(2) Think about the causes of vulnerability in people with disabilities and identify resources in your community that help build resilience.

(3) Consider the factors that lead to individual resilience. Craft an intervention that the emergency management and public health entities in your locality could implement to promote individual resilience in the community.

References

[1] Phe.gov., 2019. Available from: https://www.phe.gov/Preparedness/planning/abc/Pages/community-resilience.aspx.

[2] Acosta J, Chandra A, Madrigano J. An agenda to advance integrative resilience research and practice key themes from a resilience roundtable, 2017. [cited 2022 Jan 10]. Available from: https://www.rand.org/content/dam/rand/pubs/research_reports/RR1600/RR1683/RAND_RR1683.pdf.

[3] Fingerle M. Current state of research regarding the resilience concept (Aktueller Forschungsstand zum Resilienzkonzept). Jugendhilfe 2009;47(3):204–8.

[4] Bradford K, Garcia A. Adverse Childhood Experiences [Internet]. www.ncsl.org. 2021. Available from: https://www.ncsl.org/research/health/adverse-childhood-experiences-aces.aspx.

[5] IFRC. What is a disaster? IFRC; 2019. Available from: https://www.ifrc.org/en/what-we-do/disaster-management/about-disasters/what-is-a-disaster/.

[6] Ausmussan K, Fischer F, Drayton E, McBride T. Adverse childhood experiences: What we know, what we don't know, and what should happen next. Early Intervention Foundation; 2020. Available from: https://www.eif.org.uk/report/adverse-childhood-experiences-what-we-know-what-we-dont-know-and-what-should-happen-next.

[7] CDC. About the CDC-Kaiser ACE study |Violence prevention. Injury Center | CDC; 2020. Available from: https://www.cdc.gov/violenceprevention/aces/about.html.

[8] Felitti VJ, Anda RF, Nordenberg D, Williamson DF, Spitz AM, Edwards V, et al. Relationship of childhood abuse and household dysfunction to many of the leading causes of death in adults: the adverse childhood experiences (ACE) study. Am J Prev Med 2019;56(6):774–86. Available from: https://www.sciencedirect.com/science/article/pii/S0749379719301436.

[9] Jones CM, Merrick MT, Houry DE. Identifying and preventing adverse childhood experiences. JAMA 2019. [cited 2019 Nov 8]; Available from: https://jamanetwork.com/journals/jama/fullarticle/2755266.

[10] Office of Head Start | ACF. Home, 2019. Available from: https://www.acf.hhs.gov/ohs.

[11] Schanzenbach DW, Bauer L. The long-term impact of the Head Start program. Brookings; 2016. Available from: https://www.brookings.edu/research/the-long-term-impact-of-the-head-start-program/.

[12] Hoynes H, Schanzenbach DW, Almond D. Long-run impacts of childhood access to the safety net. Am Econ Rev 2016;106(4):903–34. Available from: https://gspp.berkeley.edu/assets/uploads/research/pdf/Hoynes-Schanzenbach-Almond-AER-2016.pdf.

[13] Kaminer D, Seedat S, Stein DJ. Post-traumatic stress disorder in children. World Psychiatry 2005;4(2):121–5. Available from: https://www.ncbi.nlm.nih.gov/pmc/articles/PMC1414752/.

[14] American Psychiatric Association. Diagnostic and statistical manual of mental disorders. 5th ed. Arlington, VA: American Psychiatric Association; 2013.

[15] Osofsky J, Kronenberg M, Bocknek E, Cross HT. Longitudinal impact of attachment-related risk and exposure to trauma among young children after hurricane Katrina. Child Youth Care Forum 2015;44(4):493–510.

[16] King LS, Osofsky JD, Osofsky HJ, Weems CF, Hansel TC, Fassnacht GM. Perceptions of trauma and loss among children and adolescents exposed to disasters a mixed-methods study. Curr Psychol 2015;34(3):524–36.

[17] Mclaughlin KA, Fairbank JA, Gruber MJ, Jones RT, Lakoma MD, Pfefferbaum B, et al. Serious emotional disturbance among youths exposed to hurricane Katrina 2 years postdisaster. J Am Acad Child Adolesc Psychiatry 2009;48(11):1069–78.

[18] Hafstad GS, Haavind H, Jensen TK. Parenting after a natural disaster: a qualitative study of Norwegian families surviving the 2004 tsunami in Southeast Asia. J Child Fam Stud 2011;21(2):293–302.

[19] Juth V, Silver RC, Seyle DC, Widyatmoko CS, Tan ET. Post-disaster mental health among parent–child dyads after a major earthquake in Indonesia. J Abnorm Child Psychol 2015;43(7):1309–18.

[20] Proctor LJ, Fauchier A, Oliver PH, Ramos MC, Rios MA, Margolin G. Family context and young children's responses to earthquake. J Child Psychol Psychiatry 2007;48(9):941–9.

[21] Hardin SB, Weinrich M, Garrison C, Weinrich S, Hardin TL. Psychological distress of adolescents exposed to Hurricane Hugo. J Trauma Stress 1994;7(3):427–40.

[22] Giannopoulou I, Strouthos M, Smith P, Dikaiakou A, Galanopoulou V, Yule W. Post-traumatic stress reactions of children and adolescents exposed to the Athens 1999 earthquake. Eur Psychiatry 2006;21(3):160–6.

[23] Goenjian AK, Molina L, Steinberg AM, Fairbanks LA, Alvarez ML, Goenjian HA, et al. Posttraumatic stress and depressive reactions among Nicaraguan adolescents after Hurricane Mitch. Am J Psychiatry 2001;158(5):788–94.

[24] Jensen TK, Dyb G, Nygaard E. A longitudinal study of posttraumatic stress reactions in Norwegian children and adolescents exposed to the 2004 tsunami. Arch Pediatr Adolesc Med 2009;163(9):856.

[25] Chemtob CM, Nomura Y, Rajendran K, Yehuda R, Schwartz D, Abramovitz R. Impact of maternal posttraumatic stress disorder and depression following exposure to the September 11 attacks on preschool children's behavior. Child Dev 2010;81(4):1129–41. [cited 2019 Mar 9] Available from: https://www.ncbi.nlm.nih.gov/pmc/articles/PMC3124807/.

[26] Menendez D, Hernandez IG, Rosengren KS. Children's emerging understanding of death. Child Dev Perspect 2020;14(1):55–60.

[27] Deering CG. A cognitive developmental approach to understanding how children cope with disasters. J Child Adolesc Psychiatr Nurs 2000;13(1):7–16.

[28] Catani C, Gewirtz AH, Wieling E, Schauer E, Elbert T, Tsunami NF. War, and cumulative risk in the lives of Sri Lankan Schoolchildren. Child Dev 2010;81(4):1176–91. [cited 2020 Jan 10]. Available from: https://onlinelibrary.wiley.com/doi/10.1111/j.1467-8624.2010.01461.x.

[29] Kronenberg ME, Hansel TC, Brennan AM, Osofsky HJ, Osofsky JD, Lawrason B. Children of Katrina: lessons learned about postdisaster symptoms and recovery patterns. Child Dev 2010;81(4):1241–59.

[30] Cloitre M, Stolbach BC, Herman JL, van der Kolk B, Pynoos R, Wang J, et al. A developmental approach to complex PTSD: childhood and adult cumulative trauma as predictors of symptom complexity. J Trauma Stress 2009;22(5):399–408. [cited 2019 Apr 7]. Available from: https://onlinelibrary.wiley.com/doi/full/10.1002/jts.20444.

[31] Shonkoff JP, Garner AS, Siegel BS, Dobbins MI, Earls MF, Garner AS, et al. The lifelong effects of early childhood adversity and toxic stress. Pediatrics 2011;129(1):e232–46. Available from: https://pediatrics.aappublications.org/content/pediatrics/129/1/e232.full.pdf.

[32] Masten AS, Cicchetti D. Developmental cascades. Dev Psychopathol 2010;22(3):491–5. Mullet-Hume E, Anshel D, Guevara V, Cloitre M. Cumulative trauma and posttraumatic stress disorder among children exposed to the 9/11 World Trade Center attack. Am J Orthopsychiatry 2008;78(1):103–8.

[33] Bonanno GA, Mancini AD. The human capacity to thrive in the face of potential trauma. Pediatrics 2008;121(2):369–75.

[34] Mullet-Hume E, Guevara V, Anshel D, Cloitre M. Cumulative trauma and posttraumatic stress disorder among children exposed to the 9/11 World Trade Center attack. Am J Orthopsych 2008;78(1):103–8.

[35] Bokszczanin A. PTSD symptoms in children and adolescents 28 months after a flood: age and gender differences. J Trauma Stress 2007;20(3):347–51 [PubMed: 17598138].

[36] Goenjian AK, Walling D, Steinberg AM, Karayan I, Najarian LM, Pynoos R. A prospective study of posttraumatic stress and depressive reactions among treated and untreated adolescents 5 years after a catastrophic disaster. Am J Psychiatry 2005;162(12):2302–8.

[37] Baddam John P, Russell S, Russell PSS. The prevalence of posttraumatic stress disorder among children and adolescents affected by tsunami disaster in Tamil Nadu. Disaster Manag Response 2007;5(1):3–7.

[38] Self-Brown S, Lai BS, Harbin S, Kelley ML. Maternal posttraumatic stress disorder symptom trajectories following hurricane Katrina: an initial examination of the impact of maternal trajectories on the well-being of disaster-exposed youth. Int J Public Health 2014;59(6):957–65.

[39] WorldAtlas. The US states most prone to natural disasters. WorldAtlas; 2018. Available from: https://www.worldatlas.com/articles/the-10-states-most-prone-to-natural-disasters.html.

[40] Furr JM, Comer JS, Edmunds JM, Kendall PC. Disasters and youth: a meta-analytic examination of posttraumatic stress. J Consult Clin Psychol 2010;78(6):765–80. Available from: https://www.ncbi.nlm.nih.gov/pubmed/21114340.

[41] Masten AS. Ordinary magic: Resilience in development. New York, NY: The Guilford Press; 2015.

[42] Weems CF, Graham RA. Resilience and trajectories of posttraumatic stress among youth exposed to disaster. J Child Adolesc Psychopharmacol 2014;24(1):2–8.

[43] Weems CF, Scott BG, Banks DM, Graham RA. Is T.V. traumatic for all youths? The role of preexisting posttraumatic stress symptoms in the link between disaster coverage and stress. Psychol Sci 2012;23(11):1293–7. https://doi.org/10.1177/0956797612446952.

[44] Couch SR, Coles CJ. Community stress, psychosocial hazards, and EPA decision-making in communities impacted by chronic technological disasters. Am J Public Health 2011;101(Suppl 1):S140–8. [cited 2020 Jun 11]. Available from: https://www.ncbi.nlm.nih.gov/pmc/articles/PMC3222505/.

[45] Bronfenbrenner U. Developmental research, public policy, and the ecology of childhood. Child Dev 1974;45(1):1.

[46] Weems CF, Overstreet S. Child and adolescent mental health research in the context of Hurricane Katrina: an ecological needs-based perspective and introduction to the special section. J Clin Child Adolesc Psychol 2008;37(3):487–94. Guy-Evans O. Bronfenbrenner's ecological systems theory. Simply Psychology; Nov 09, 2020. https://www.simplypsychology.org/Bronfenbrenner.html.

[47] Ohmer ML. How theory and research inform citizen participation in poor communities: the ecological perspective and theories on self- and collective efficacy and sense of community. J Hum Behav Soc Environ 2010;20(1):1–19.

[48] Kilmer RP. Helping families and communities recover from disaster: lessons learned from Hurricane Katrina and its aftermath., 2010, Washington, DC: American Psychological Association.

[49] Scaramella LV, Sohr-Preston SL, Callahan KL, Mirabile SP. A test of the family stress model on toddler-aged children's adjustment among Hurricane Katrina impacted and nonimpacted low-income families. J Clin Child Adolesc Psychol 2008;37(3):530–41. [cited 2019 Sep 4]. Available from: https://www.ncbi.nlm.nih.gov/pmc/articles/PMC2893557/.

[50] Kerns CE, Elkins RM, Carpenter AL, Chou T, Green JG, Comer JS. Caregiver distress, shared traumatic exposure, and child adjustment among area youth following the 2013 Boston Marathon bombing. J Affect Disord 2014;167:50–5.

[51] Aldrich DP, Meyer MA. Social capital and community resilience. Am Behav Sci 2014;59(2):254–69.

[52] Fernandez A, Black J, Jones M, Wilson L, Salvador-Carulla L, Astell-Burt T, et al. Flooding and mental health: a systematic mapping review. Ebi KL, (ed), PLoS ONE 2015;10(4), e0119929. Available from: https://journals.plos.org/plosone/article%3Fid%3D10.1371/journal.pone.0119929.

[53] Boscarino JA, Hoffman SN, Kirchner HL, Erlich PM, Adams FE, Figley CR, et al. Mental health outcomes after Hurricane Sandy. Int J Emerg Ment Health 2015;15(3):147–58.

[54] Paxson C, Fussell E, Rhodes J, Waters M. Five years later: recovery from post traumatic stress and psychological distress among low-income mothers affected by hurricane Katrina. Soc Sci Med 2012;74(2):150–7.

[55] Beinecke R, Raymond A, Cisse M, Renna K, Khan S, Fuller A, et al. The mental health response to the Boston bombing: a three-year review. Int J Ment Health 2017;46(2):89–124.

[56] Garfin DR, Holman EA, Silver RC. Exposure to prior negative life events and responses to the Boston marathon bombings. In: Psychological trauma: Theory, research, practice, and policy; 2019 Sep 26.

[57] Hobfoll SE, Canetti-Nisim D, Johnson RJ, Palmieri PA, Varley JD, Galea S. The association of exposure, risk, and resiliency factors with PTSD among Jews and Arabs exposed to repeated acts of terrorism in Israel. J Trauma Stress 2008;21(1):9–21.

[58] Höfler M. Psychological resilience building in disaster risk reduction: contributions from adult education. Int J Disaster Risk Sci 2014;5(1):33–40.

[59] Peter G, Van der Velden C, Yzermans J, Grievink L. Enschede fireworks disaster. In: Neria Y, Galea S, Norris FH, editors. Mental health and disasters. Cambridge University Press; 2009. p. 473–96.

[60] Lowe SR, Rhodes JE, Waters MC. Understanding resilience and other trajectories of psychological distress: a mixed-methods study of low-income mothers who survived hurricane Katrina. Curr Psychol 2015;34(3):537–50.

[61] Lai BS, Tiwari A, Beaulieu BA, Self-Brown S, Kelley ML. Hurricane Katrina: maternal depression trajectories and child outcomes. Curr Psychol 2015;34(3):515–23.

[62] Kessler RC, Barker PR, Colpe LJ, Epstein JF, Gfroerer JC, Hiripi E, et al. Screening for serious mental illness in the general population. Arch Gen Psychiatry 2003;60(2):184.

[63] Jonkman SN, Kelman I. An analysis of the causes and circumstances of flood disaster deaths. Disasters 2005;29(1):75–97.

[64] Stough LM. The effects of disaster on the mental health of individuals with disabilities, 2009. [cited 2022 Jan 16]. Available from: https://oaktrust.library.tamu.edu/bitstream/handle/1969.1/153544/The%20effects%20of%20disaster%20on%20the%20mental%20health%20of%20individuals%20with%20disabilities.pdf?sequence=1.

[65] Harvard School of Public Health. Hurricane readiness in high-risk areas, 2007. [cited 2022 Jan 16]. Available from: https://www.hsph.harvard.edu/news/press-releases-search/.

[66] Brodie M, Weltzien E, Altman D, Blendon RJ, Benson JM. Experiences of hurricane Katrina evacuees in Houston shelters: implications for future planning. Am J Public Health 2006;96(8):1402–8.

[67] McGuire LC, Ford ES, Okoro CA. Natural disasters and older US adults with disabilities: implications for evacuation. Disasters 2007;31(1):49–56.

[68] Rooney C, White GW. Consumer perspective. J Disabil Policy Stud 2007;17(4):206–15.

[69] Ballan MS, Sormanti M. Trauma, grief and the social model: practice guidelines for working with adults with intellectual disabilities in the wake of disasters. Rev Disabil Stud 2006;2(3). [cited 2022 Jan 16]. Available from: https://www.rdsjournal.org/index.php/journal/article/view/339.

[70] Valenti-Hein D, Schwartz LD. The sexual abuse interview for those with developmental disabilities. Santa Barbara CA: James Stanfield Co., Inc; 1995.

[71] Takahashi A, Watanabe K, Oshima M, Shimada H, Ozawa A. The effect of the disaster caused by the great Hanshin earthquake on people with intellectual disability. J Intellect Disabil Res 1997;41(2):193–6.

[72] Bonanno GA, Galea S, Bucciarelli A, Vlahov D. What predicts psychological resilience after disaster? The role of demographics, resources, and life stress. J Consult Clin Psychol 2007;75(5):671–82.

[73] Seery MD. Resilience. Curr Dir Psychol Sci 2011;20(6):390–4.

[74] Seery MD, Holman EA, Silver RC. Whatever does not kill us: cumulative lifetime adversity, vulnerability, and resilience. J Pers Soc Psychol 2010;99(6):1025–41.

[75] Rodriguez-Llanes JM, Vos F, Guha-Sapir D. Measuring psychological resilience to disasters: are evidence-based indicators an achievable goal? Environ Health 2013;12(1).

[76] Sherman MF, Gershon RR, Riley HEM, Zhi Q, Magda LA, Peyrot M. Emergency preparedness safety climate and other factors associated with mental health outcomes among world trade center disaster evacuees. Disaster Med Public Health Prep 2016;11(3):326–36.

The US emergency management and public health preparedness systems

Susan Wolf-Fordham[a,b]

[a]*Consultant, Association of University Centers on Disabilities, Silver Spring, MD, United States,* [b]*Adjunct Faculty of Public Health, Massachusetts College of Pharmacy and Health Sciences, Boston, MA, United States*

Learning objectives

After completing this chapter, you will be able to:

- Describe the components, functions, and frameworks of the federal, state and local emergency systems
- Explain the impact of federalism and the US Constitution on the emergency system
- Explain how attention to disability and other functional and access need issues fits and does not fit within the emergency system

Introduction

In order to provide a context for considering the impact of emergencies and disasters on individuals with disabilities and strategies for increasing access and equity, it is important to understand the US emergency management and public health preparedness systems (together, the emergency system). This chapter describes how the current emergency system evolved from fragmentation and disorganization to the still somewhat siloed "system of systems" that it is today, with parallel public health and public safety/emergency management systems at the local (city, town, county), state, territorial, or tribal, and federal levels. All response begins at the local level and then moves up through the state and federal levels, and the bottom up nature of the system is explained. Authority is diffuse throughout the emergency system. The Federal Emergency Management Agency (FEMA) does not control the state emergency management system and the US Department of Health and Human Services (HHS) (with offices responsible for public health and healthcare preparedness and response) does not control state public health agencies. In many states the local public health and emergency management systems are independent of the state systems. The chapter describes the federal emergency declaration process and the "all hazards" integrated approach which focuses on commonalities among risks rather than a single particular risk. Key actors and their roles within the federal, state and local emergency systems are discussed as well as the guiding principles that apply to the emergency lifecycle phases of mitigation-preparedness-response-recovery. One set of guiding principles, the National Incident Management System (NIMS), is operationalized in part through the Incident Command System (ICS), a management structure for emergency response which is described below. The chapter examines two features at the local level, the level closest to the people likely to be most directly impacted by a disaster: registries (databases with the names, addresses, and phone numbers of people with disabilities) and medical shelters (disaster shelters separate from those for the general public). Structural challenges related to system silos are discussed, as well as factors that have led to the exclusion of disabled people from planning and response processes. The chapter also considers the FEMA and HHS departments responsible for civil rights compliance and enforcement related to emergency systems.

Evolution of the American emergency system

Historically the United States' response to disasters and emergencies was not systematized; central direction did not occur until the 1970s. Some of the first examples of disaster response organization include Benjamin Franklin's creation of a volunteer fire company in Philadelphia in 1736 to promote best practices in firefighting [1] and Clara Barton's creation of a volunteer group in 1881, later to become the American Red Cross, to address the human consequences of personal and large scale disasters [2]. However, until the establishment of presidential authority to issue disaster declarations and provide federal disaster support through the Disaster Relief Act of 1950, and the creation of the Office of Emergency Preparedness inside the White House in 1961, most response to disasters was fragmented, local, and spontaneous [3,4].

By the 1970s more than 100 agencies existed with responsibility for some aspect of risk management and disaster response activities for various types of disasters. This fragmentation created inefficiencies and duplication of effort as well as power struggles and confusion over which entity held responsibility for which activities. During the 1970s the National Governor's Association, among others, called for a national focus on emergency management with one federal agency in the lead [3].

In response to these calls to action, in 1979 the Federal Emergency Management Agency (FEMA), a federal government agency, was created in an attempt to create order and decrease the confusion and fragmentation. After the September 11, 2001 terror attacks, FEMA moved into the newly created federal Department of Homeland Security (DHS) [3].

While public health has been an important government function since well before the American Public Health Association was founded in 1872 [5], the public health preparedness field is much newer. Public health preparedness programs began in the late 1990s and originally focused on planning for potential bioterrorism attacks [6]. Due to the September 11, 2001 attacks, Congress increased public health infrastructure funding and public health began to be thought of as including an emergency response function. In 2006, after the injuries, deaths, displacement and other health consequences of Hurricane Katrina (2005), Congress created the Office of the Assistant Secretary for Preparedness and Response (ASPR) in the US Department of Health and Human Services, and the public health preparedness field broadened its focus to include the health aspects of diverse emergencies, including natural, accidental, and human caused disasters and pandemics [6].

The US emergency system has evolved from a fragmented system to a more structured system, with two parallel components: public health preparedness and emergency management. The sections below discuss the principles and frameworks undergirding the current emergency system, the "all hazards" model of focusing on commonalities among different incidents and preparedness for a broad range of disasters, the disaster life cycle, the structure of the parallel systems at the local, state, tribal and federal levels, and the importance of federal disaster declarations.

Guiding principles and frameworks
The National Preparedness Goal and guiding principles for response

The National Preparedness Goal [7] and the National Response Framework [8] guide current US emergency planning and response. The National Preparedness Goal defines preparedness across all emergency systems, and is succinct: "A secure and resilient nation with the capabilities required across the whole community to prevent, protect against, mitigate, respond to, and recover from the threats and hazards that pose the greatest risk" [7, p. 1]. The risks include natural, accidental, and human caused disasters and pandemics. The Goal encompasses 32 capabilities which describe the skills, knowledge, and competencies the whole community needs to achieve the preparedness goal. The National Response Framework contains the current guiding principles, roles and structures under which the country responds to emergencies and disasters, rather than acting as a specific response plan for the country. The Framework incorporates the concept that responsibility for an effective disaster response is shared between all levels of government, plus stakeholders in the public and private sectors, communities and citizens, which would encompass all populations, including people with disabilities and others with access and functional needs [8]. In addition to a goal and guiding principles, there is a system for coordination, information sharing and interoperability (ensuring that technologies or other systems communicate with each other and work seamlessly together) within the system.

Coordination, information sharing and interoperability

Efficient and effective emergency management and public health preparedness require sound coordination, an effective information sharing system, seamless technology and other interoperability among governments and other stakeholders. This becomes particularly important where, as in the United States, there are multiple systems at the federal, state and local levels. The US framework to accomplish these aims is the National Incident Management System (NIMS), a standardized, comprehensive national coordination framework which guides emergency incident management [6,9]. The goal of NIMS is to provide a common language and structure when responding to all kinds of emergencies and disasters in all jurisdictions and at all levels. NIMS also guides how the systems, people and organizations within them work together in all phases of an emergency [10].

The Incident Command System (ICS), a part of NIMS, is the structural framework that assigns roles and responsibilities to the emergency systems' actors as they respond to emergency incidents [3,6,9,10]. Both emergency management and public health use this system; healthcare uses a modified version of ICS [6]. Under ICS there is a clear chain of command and accountability with explicit roles and responsibilities for response [10,11]. ICS organizes response personnel into three groups. The Incident Commander leads the response, and is supported by Incident Command Staff such as the Public Information, Safety, and Liaison Officers, and General Staff related to Operations, Planning, Logistics and Finance/Administration.

Within ICS, the Public Information Officers advise the Incident Commander regarding communication and information sharing with the media and the public. The Safety Officer provides advice about responder safety and briefs responders about safety issues. The Liaison Officer is the communication channel between different entities supporting the response [3,10,11]. The General Staff includes four Sections: (1) Operations, (2) Planning, (3) Logistics, and (4) Finance/Administration. The Operations Section has responsibility for response operations (e.g., evacuation). The Planning Section collects, evaluates, and shares intelligence, maintains key documents related to the response and prepares the Incident Action Plan with high level goals to address the incident. The Logistics Section directs communications, medical and food services for responders, and obtains supplies and resources needed for the response. The Finance/Administration Section maintains personnel records and contracts, and conducts cost analyses [3,10,11]. Each section contains Emergency Support Functions (ESF), or areas of responsibility to be undertaken by a government agency assigned to be part of an emergency response. ESFs are referred to by number. For example, in a state ICS system, the Operations Section oversees the state public health department which is responsible for medical care and public health services (ESF 8) during an emergency. The responsibility for mass care (disaster sheltering), part of ESF 6, is often assigned to the American Red Cross, under state public health agency supervision [3,10,11]. Fig. 1 illustrates a basic ICS structure for an incident in one jurisdiction.

In a response to a small, localized natural disaster the local emergency manager would likely coordinate the Emergency Operations Center; the on-scene Incident Commander would depend on the nature of the incident. For example, in the case of a fire the local Fire Chief might be the Incident Commander, the head of the local government's communications office the Public Information Officer, and the police chief the Safety Officer [10,11]. The local public health and human services department might be in charge of emergency shelters, temporary housing, human services, medical and public health services [8]. When an emergency incident is more complex, such as when more than one government agency has jurisdiction or the incident happens across jurisdictions, modifications to the ICS

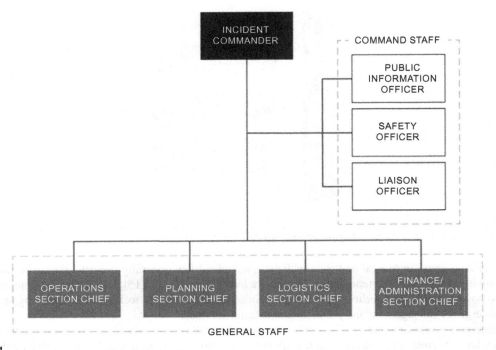

FIG. 1

The Incident Command System organizational chart, with leaders for a typical single jurisdiction emergency.

Credit: United States Department of Homeland Security FEMA. National incident management system. 3rd ed.
United States Department of Homeland Security/FEMA; 2017. Available from: https://www.fema.gov/sites/default/files/2020-07/
fema_nims_doctrine-2017.pdf.

system are made to set coordinated objectives and strategies for addressing the emergency, with clear delineation of the responsibilities shared among the actors involved [3,10,11].

The US emergency system structure

While some countries, such as Israel and Cuba, have a single national emergency system, the United States does not, as noted above. Rather, the US emergency system is essentially divided into six separate but interrelated systems. Distinct emergency management and public health preparedness systems exist at the local, state, regional, and federal levels as shown in Fig. 2.

The US government is based on a federalism model, which means that the federal, state and local levels of government have separate powers and authority, including powers and authority related to disaster planning and response. The decentralization of the US emergency system reflects historic American views of politics and traditional distrust of central authority. This distrust led the Founding Fathers to design a governmental structure where the federal, state, and local governments share power, with each generally having its own areas of authority [12,13]. Under the US Constitution, where there

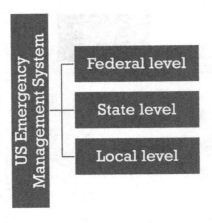

FIG. 2

The three governmental levels of the US emergency management system.

are conflicts between federal and state or local law, federal law generally prevails. However, powers that are not reserved by the federal government are left to the states [14,15]. Consequently, states retain broad powers and responsibility to protect the public health, safety and welfare of residents by providing emergency management and public health disaster services.

Emergency management professionals often say that "all response is local." Based on federalism principles, US emergency planning and response is a bottom-up system, that begins at the local city, town, or county level [16]. State or territorial intervention generally comes only at the request of the local emergency management system when the locality determines that its emergency resources are overwhelmed. Likewise, the national emergency management agency, FEMA, generally cannot intercede in a state's emergency unless the state's governor makes a specific request to the US President [3]. Once a Presidential disaster declaration has been made, federal disaster financial and other assistance becomes available to a state, territory, or tribal government. Federal resources supplement, but don't replace or take control of, local and state efforts and resources [16].

A key concept supporting the US emergency system is the "all-hazards" model [17]. This model is based on the premise that generic policies, processes and activities can be applied to different types of disasters. For example, hurricanes or other weather-related events and cyber security breaches of the electrical grid can cause power outages. Preparing for and responding to power outages, whatever the cause, requires similar steps, and lessons learned during a power outage due to one kind of emergency are applicable to a power outage caused by another event.

The disaster life cycle and phases

Another important concept is the disaster life cycle with its four emergency management phases. The cycle has no clear beginning or end; rather it is continuous and encompasses the preparedness, mitigation, response and recovery phases of a disaster event [18]. Mitigation refers to reducing or eliminating structural or other risks such as building dams to prevent flooding prior to hurricane season. Preparedness involves training, planning, or emergency plan testing via different kinds of exercises or drills [17]. Response consists of immediate activities taken during an emergency event to save life,

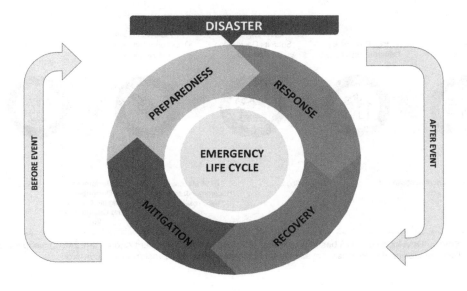

FIG. 3

The disaster life cycle, with four phases.

property, or sustain well-being, such as evacuation, providing medical care, or opening emergency shelters. During the recovery phase after an emergency, activities targeted at helping return communities and their members back to normal, such as restoring electricity, reopening government services, and building new housing occur [17]. Building resilience may be considered within each part of the cycle, with the ultimate goal of strengthening individuals and communities in order to minimize the impact of future emergencies. Fig. 3 is an image of the continuous disaster life cycle, showing the four phases.

The four phases of emergency management are not necessarily distinct and may have considerable operational overlap. Rebuilding a school post-disaster, for example, is part of the recovery phase. Rebuilding a school to withstand high winds that cause damage may be both recovery and mitigation. Some authorities add a fifth phase, "prevention," to the cycle. Prevention activities diminish vulnerability and may prevent or minimize the potential of future harm to individuals and communities [19]. For example, sound environmental planning and building design standards may minimize a disaster's impact.

Disaster declarations and recovery funding

The United States has a bottom up emergency system in which federal support comes *only* after first local, and then state, resources become overwhelmed. Under the federal Stafford Act the President must declare an emergency, major disaster, or fire declaration in order for states, territories, tribes and localities to access federal disaster funding [16]. Generally, the Governor, territorial leader, or Tribal Chief Executive must specifically request that the President make such a declaration and then the request is reviewed. FEMA provides a recommendation based on a preliminary damage assessment, but the President is solely responsible for making the decision about a declaration. In limited circumstances

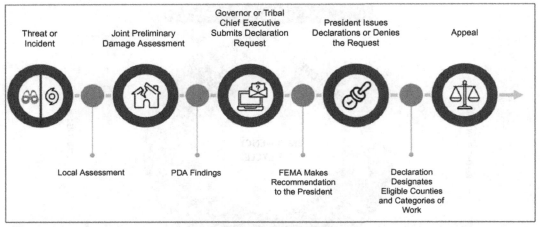

Source:Developed by CRS based on the Fedral Emergency Management Agency's (FEMA's) webpage on "The Disaster Declaration Process, " available at https://www.fema.gov/disaster-declaration-process.

FIG. 4

Stafford act presidential emergency declaration process.

Credit: Congressional Research Service. Congressional primer on responding to and recovering from major disasters and emergencies. Congressional Research Service; Update June 3 2020 [cited 2021 Dec 29]. Available from: https://sgp.fas.org/crs/homesec/R41981.pdf: p. 9.

the President may act unilaterally, which is what happened in 2019 when the President declared the COVID-19 pandemic to be a disaster. Fig. 4 shows the Stafford Act emergency declaration process.

Not all disasters or emergencies are federally declared and therefore eligible for federal funding. FEMA only provides financial assistance to states, tribes, territories, localities and certain non-profit organizations for federally declared disasters. The agency can provide financial assistance just before or after an emergency and for long term recovery. Between 2011 and March 2021 FEMA provided financial assistance in every state and territory [20,21]. FEMA funding involves a cost share between FEMA and the state where the disaster occurred [20]. FEMA also provides financial assistance to individuals and families affected by disaster who are eligible and uninsured or underinsured [22]. This assistance may include funding for temporary housing or to support home repair or replacement. Funds for other expenses due to an emergency or disaster, such as medical care, childcare, and funeral costs may also be covered. This financial assistance does not cover the cost of all losses; federal funding supplements, rather than replaces, other available funding [22].

A system of systems

The US emergency system is really "a system of systems" [23, p. 2]. As noted earlier in this chapter, the federal nature of the US government has led to emergency management and public health preparedness systems at the local, state and federal levels. Within each government level, *both* a public safety/emergency management system and a separate public health preparedness system exist, creating two parallel systems. Each of these, the emergency management system and the public health system, are

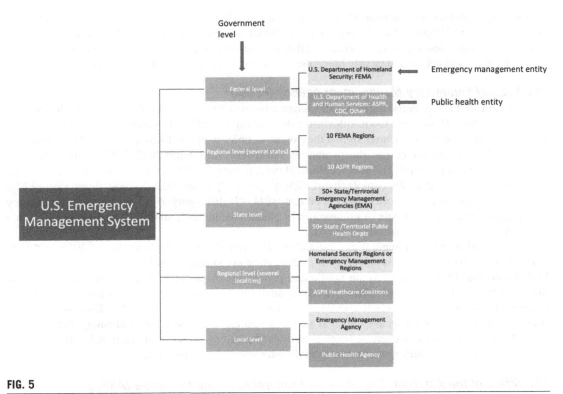

FIG. 5

The US Emergency Management "System of Systems."

present at the federal level, the state and territorial level (50 states plus all territories), and the county and/or city level. Counties, cities and states have also partnered to set up regional systems within and among states. Fig. 5 shows this system of systems.

There is a saying in the field that disaster response is "locally executed, state managed, federally supported" [24] which aptly describes the US system of systems. The following sections of this chapter take a closer look at emergency system components by governmental level, and describe key emergency management and public health preparedness agencies.

The national level: The Federal Emergency Management Agency (FEMA), Office of the Assistant Secretary for Preparedness and Response (ASPR), and the Centers for Disease Control and Prevention (CDC)

The two main federal agencies responsible for emergency management and public health preparedness planning and response at the national level are the Federal Emergency Management Agency (FEMA) and the Office of the Assistant Secretary for Preparedness and Response (ASPR). As the federal level emergency management organization, FEMA is housed within the US Department of Homeland Security. FEMA "coordinates the federal government's role in preparing for, preventing, mitigating the effects of, responding to, and recovering from all domestic disasters, whether natural or man-made, including acts of terror" [25]. ASPR sits in the US Department of Health and Human Services (HHS) and manages the planning, response, and recovery activities of HHS to ensure the integration of public and environmental health, medical and behavioral health, human services, and responder health and safety [11,26].

The Federal Emergency Management Agency (FEMA)

The concept of FEMA dates back to 1803 when Congress enacted a statute to provide disaster assistance to a New Hampshire town following a devastating fire [27]. However, this didn't become official until President Jimmy Carter signed the executive order to create FEMA in 1979. The mission of FEMA is "helping people before, during and after disasters" [28]. Until 2022 FEMA implemented its mission through three aims: (1) "build a culture of preparedness"; (2) "ready the nation for catastrophic disasters"; and (3) "reduce the complexity of FEMA" [29, p. 3]. In its 2022–2026 Strategic Plan FEMA adjusted the goals to place more emphasis on equity: (1) "Instill equity as a foundation for emergency management"; (2) "Lead whole of community in climate resilience"; and (3) "Promote and sustain a ready FEMA and prepared nation" [30,p. 3].

FEMA's 2018–2022 strategic plan [29] and the Disaster Recovery Reform Act of 2018 [31] both acknowledge that the sole responsibility for emergency preparedness and response cannot rest with FEMA and that FEMA must rely on public and private sector partners to accomplish its goals. The 2022–2026 Strategic Plan emphasizes shared responsibility for emergency preparedness and response, community partnerships and whole community outcomes even more strongly [29]. This partnership is necessary as FEMA does not have the legal authority to direct either state or local emergency management. However, FEMA can exercise some control over state emergency management by defining how the federal agency provides grant funding and disaster assistance to the states.

The Office of the Assistant Secretary for Preparedness and Response (ASPR)

The other agency at the national level with responsibility related to emergency and disaster planning and response, the US Department of Health and Human Services' Office of the Assistant Secretary for Preparedness and Response (ASPR), is much newer than FEMA. It was created in 2006 in the wake of Hurricane Katrina. Under the Pandemic and All Hazards Preparedness Act of 2006ASPR was charged with heading the nation's response in the event of a public health disaster [32] ASPR's mission and vision are:

> ASPR leads the nation's medical and public health preparedness for, response to, and recovery from disasters and public health emergencies. ASPR collaborates with hospitals, healthcare coalitions, biotech firms, community members, state, local, tribal, and territorial governments, and other partners across the country to improve readiness and response capabilities [33].

ASPR focuses on: (1) preparedness planning and response; (2) building federal emergency medical operational capabilities; (3) medical countermeasure research; (4) and development, procurement, and maintenance of the Strategic National Stockpile. Medical countermeasures refer to drugs, including vaccines, and medical supplies or devices used to address potential public health emergencies. The Strategic National Stockpile includes medical countermeasures that are stored by the federal government. The agency also provides grants to strengthen the emergency, disaster, and pandemic capabilities of hospitals and health care systems [33,34].

ASPR's *Health Care Preparedness and Response Capabilities* [35], outlines broad, high level preparedness, response and recovery goals for the US healthcare delivery system, including, among others, hospitals, Emergency Medical Service (EMS) organizations, home health agencies, health care professional organizations, primary care providers and outpatient health care centers. The office also provides federal support, including medical professionals, through ASPR's National Disaster Medical System, to augment state and local capabilities during an emergency or disaster [33]. The ASPR Healthcare Coalition (HCC) program [36] is of particular note. The coalitions are organized regionally within states, and consist of hospitals, emergency medical services (EMS), and emergency management agencies, plus public health agencies in a defined geographic area [35]. HCCs act as coordinating groups and provide support to health care system emergency preparedness and response, particularly relating to the medical care and public health emergency support function (ESF 8) [35,36]. Like FEMA, ASPR does not have the legal authority to direct state or local public health departments, but develops guidance for the field and incentivizes action via grantmaking authority.

How FEMA and ASPR address the emergency needs and civil rights of people with disabilities

Both FEMA and ASPR address the emergency needs of people with disabilities through offices tasked with assuring that these needs are considered and met. FEMA hired its first Disability Coordinator in 2007, as required by the Post-Katrina Emergency Management Reform Act of 2006 [37,38]. Since 2010, the FEMA Office of Disability Integration and Coordination (ODIC), has actively engaged state and local governments, as well as public and not for profit organizations, to provide them with tools to foster equal access and delivery of disaster services to people with disabilities throughout the emergency management cycle. FEMA's ODIC also holds disability stakeholder calls and convenes other meetings [38]. Regional disability integration advisors with expertise in disability and emergency planning aid FEMA, states, territories, and communities in addressing the needs of people with disabilities. These regional advisors provide training, technical assistance, and advice to FEMA programs as well as to senior decision makers and emergency managers in the field [38]. ODIC provides guidance, resources and strategies that encourage the inclusion and participation of people with disabilities and in emergency management practices throughout the disaster cycle [38].

The law requires equal rights to emergency and disaster services and functions. FEMA's Office of Equal Rights supports equal access to FEMA programs, services and activities and its External Civil Rights Division ensures compliance with and enforcement of civil rights mandates [39]. The Office of Equal Rights addresses the common misconception that all laws can be waived during emergencies through its statement that:

Under Federal civil rights laws and the Robert T. Stafford Disaster Relief and Emergency Act (Stafford Act), FEMA, State, local, Tribal, and Territorial (SLTT) partners, and non-governmental relief and disaster assistance organizations engaged in the 'distribution of supplies, the processing of applications, and other relief and assistance activities shall [accomplish these activities] in an equitable and impartial manner, without discrimination on the grounds of race, color, religion, [national origin], sex, age, disability, English proficiency, or economic status.' Civil rights laws and legal authorities remain in effect, and cannot be waived, during emergencies [39].

Within ASPR, the Office for At-Risk Individuals, Behavioral Health and Community Resilience (ABC) bears responsibility for assuring the needs of at-risk populations (including people with disabilities and others with access and functional needs) are addressed in the medical and public health emergency planning and response process. ABC's mission is to ensure that the access and functional needs of at-risk individuals, behavioral health, and community resilience are integrated into the public health and medical emergency preparedness, response, and recovery activities of the nation by providing policy leadership, subject matter expertise to HHS and ASPR and coordination to meet the needs of those most adversely affected by disaster [40].

The use of the phrase "access and functional needs" here refers to individuals with disabilities, as well as others who may potentially experience disparate impacts from emergencies, disasters and pandemics as members of potentially at-risk groups [40].

Like FEMA, the US Department of Health and Human Services has a civil rights office, the Office for Civil Rights (OCR) [41]. The OCR is tasked with investigating complaints regarding disability discrimination. In 2021, during the COVID-19 pandemic, a number of discrimination complaints were filed against a hospital system which, due to the COVID-19 pandemic, prohibited hospitalized individuals from being accompanied by a support person [42]. The complaint alleged that the hospital system's actions denied patients with disabilities access to effective communication as required under the Americans with Disabilities Act (ADA). In response to OCR inquiries, the hospital system revised its policies to allow disabled patients to have a person they rely on for support be present with them while they are hospitalized.

The Centers for Disease Control and Prevention (CDC)

Many US Department of Health and Human Services agencies are involved in issues related to emergencies, disasters and pandemics, such as the Food and Drug Administration, the Biomedical and Advanced Research and Development Authority (BARDA), the Centers for Medicare & Medicaid Services, and the National Institutes of Health, among others. The Centers for Disease Control and Prevention (CDC), the US federal health protection agency, plays a strong role in public health preparedness [3,9,43]. The agency monitors disease outbreaks in the United States and around the world and may deploy experts onsite as needed. The CDC provides funding and technical assistance at the state and local level to promote preparedness, for example by supporting activities to build capacity, strengthen resilience and enhance incident and information management. The agency also provides resources for clinicians and non-professionals via a number of different communication channels [3,9,43]. The CDC developed extensive national standards for state, local, tribal and territorial public health agency preparedness related to such areas as incident command, community preparedness, recovery, medical material management, information sharing, medical surge, and mass care sheltering [44]. Like ASPR and FEMA guidance, CDC guidance stresses community partners, the whole community inclusion concept and the importance of addressing access and functional needs. The CDC has also developed a number of resources related to health emergencies and populations with disabilities and others with access and functional needs.[45,46].

The state/territorial level: State/territorial emergency management and public health agencies

Public health and public safety/emergency management agencies exist at the state/territorial and local levels as well. The location of these entities within the governmental organization of a state/territory may differ, but the functions assigned related to emergency planning and response generally do not.

The state public safety or emergency management agency (EMA) acts as a connector between the state and local communities [9,47]. In the current emergency system, governors have the ultimate responsibility to determine when an event reaches the level of a state disaster and requires support over and above what the local entities can provide and manage. State EMAs advise the governors about when to make a disaster declaration and request federal support once the state's emergency resources become overwhelmed [16,47]. State EMAs can also act as conduits for federal grant funds to be channeled to local communities for emergency planning, mitigation, response, or recovery. Generally state EMAs do not function as the initial first responders as that function belongs to the local emergency management and public health offices and first responders. Rather, an EMA steps in with assistance at the behest of the local emergency management agency to augment local resources [8,16,47].

In addition to designated emergency management agencies, all states have established public health agencies responsible for disease control and prevention, which often includes organized efforts between public and private organizations [44,48]. These public health entities not only bear the responsibility for activating the acute public health response in the event of an emergency, but also assessing and analyzing the potential health threats that various different disaster types bring with them. State public health departments harness their available resources such as clinical staff, laboratories, and vaccine reserves to help address the needs of the public and emergency responders related to specific disasters [44]. State public health also has a role in requesting additional supports from the CDC in order to increase state public health response capacity when needed [49].

States approach the organization of these designated agencies in different ways and the functions often sit in different places within state government. Comparing two states, Massachusetts and South Carolina, illustrates how the public health and emergency management functions are placed differently within the state government organization. The Massachusetts Emergency Management Agency (MEMA) is an agency which is part of the Executive Office of Public Safety and Security [50]. In South Carolina, on the other hand, the Emergency Management Division (SCEMD) is part of the South Carolina Department of the Military [51]. The roles and services provided by these two entities are generally similar despite their different locations. Each provides information related to personal preparedness and how to prepare for and what to do in an emergency such as a winter storm or hurricane. As well, they offer training for professionals and responders, resources for response, and information about grants and other funding sources. MEMA and SCEMD also coordinate distribution of FEMA's post-disaster individual and public assistance and disaster loans for businesses [51]. Additionally, each state entity provides guidance and information about the incorporation of people with access and functional needs into emergency management services. For South Carolina, information about self-preparedness for people with disabilities, termed people with "functional needs," is available from the SCEMD [52]. In Massachusetts, MEMA has a functional and access needs expert advisory committee and created a resource guide for local emergency management agencies to use in developing inclusive emergency plans and services [53].

Regarding public health, both Massachusetts and South Carolina public health agencies support emergency preparedness and response functions and coordinate these with other state agencies and organizations involved in emergency operations. In Massachusetts, the Department of Public Health, part of the Executive Office of Health and Human Services, has an Office of Preparedness and Emergency Management (OPEM) which functions as the source of professional planning and preparedness tools, training for professionals and information for the public, in addition to its emergency planning, response and recovery roles [54]. OPEM developed a disaster communication tool [55], a personal preparedness guide in numerous languages based on the CMIST access and functional needs framework [56], and an emergency population planning tool with data about potentially vulnerable populations such as older adults living alone, people with disabilities, and other demographic data, plus the location of community health centers and other facilities [57].

The South Carolina public health agency, the Department of Health and Environmental Control (DHEC), has responsibility for environmental protection, ocean and coastal resource management as well as public health preparedness and response to disasters [58]. DHEC provides self-preparedness information for people with functional needs [59] and a brochure about "medical special needs shelters" [60]. The criteria for housing at these shelters is unclear. DHEC provides two examples of the needs this kind of shelter may address: (1) electricity to power medical equipment; and (2) a hospital or special bed, although the brochure also states that hospital beds may not be available. The brochure further explains that people who come to the shelter should be "medically stable at home," not require hospitalization, and must bring a care giver with them [60]. There is more discussion of medical or special needs shelters below.

Medical or special needs shelters

Traditionally, many people have assumed that everyone with a disability should be housed in separate medical or special needs shelters. Diverting a disabled person to a medical shelter means separating them from friends, family, and other natural and professional supports. It may result in a loss of independence and may result in broader impacts as well. "[I]nappropriate placement can jeopardize the health and safety of the entire community by creating unnecessary surges on emergency medical resources" [61, p. 15].

FEMA guidance states that, unless someone is medically fragile, mass care (general population) shelters should meet their needs.

Children and adults with and without disabilities who have access or functional needs who require medical services may not be excluded from a general population shelter. Plans should direct that, at a minimum, medical care that can be provided in the home setting (e.g., assistance with wound management, bowel or bladder management, or the administration of medications or use of medical equipment) is available at each general population shelter [61,p. 15].

The FEMA guidance reports that the US Department of Justice calls for disabled persons to be housed in a mass care shelter even if a personal assistance service provider is not with them.

The local level: City, county and town emergency management and public health departments

US emergency response begins at the local city, town, or county level [3]. The local level is closest to the individuals and communities most impacted by an emergency, can plan in advance with local resources, and can respond more quickly than higher government levels. The local level would have the greatest familiarity with the types of disasters that occur in the area, the people who are impacted by them, and knowledge of the affected community's history, traditions and values [3]. The definition of "local" as it relates to emergency planning and response varies across the country and is dependent on the municipal organization within the state. In general, local emergency planning and response occurs at the county level, or city level for larger cities; in a small number of states, it occurs mainly at the town level [23]. The locality generally assumes the responsibility for public education about emergencies and disasters, issuing alerts and warnings, evacuation, sheltering, providing emergency medical care and developing mutual aid agreements with other communities to share resources [3,9].

There is also great variability in the size of these local planning entities even if they have the same designation. For example, in New Hampshire each city or town is responsible for its own emergency planning, even if the town has a small population. The state expects that every locality's Emergency Operation Plan meets all of the state and federal requirements. In Bedford, NH which has a population of approximately 22,000 people [62], the responsibility for local emergency planning is shared by the public health officer, the police chief, and the fire chief [63]. This model builds public safety and public health collaboration and coordination directly into the emergency system. Another state, Pennsylvania, has a law mandating that each county and municipality regardless of size develop and maintain an emergency plan [64]. Hennepin County, Minnesota, where Minneapolis is located, has a population of over 1 million people and represents an example of a large, county-based local emergency planning and response system [65]. In addition to coordinating and maintaining disaster resources in the county, Hennepin County is responsible for maintaining a partnership with each of the 45 cities in the county. Hennepin County Emergency Management responds to crises when a city emergency manager, mayor or other city official requests county resources [66]. Montana's state statute requires that each level of government be responsible for the safety and security of its residents, for keeping citizens informed about emergencies and disasters, and for providing them with assistance during emergency events [67].

Localities may have more than one emergency plan, for example a public health preparedness plan and a separate emergency plan, and maybe other plans as well. The plans may or may not integrate, coordinate or even refer to each other. Local plans may have different titles in different jurisdictions as well. While they should be publicly available, not all communities make emergency management and/or public health preparedness plans available, for example on the local government website.

Voluntary registries

Registries are computer or paper databases with the names, home addresses, and phone numbers of residents with disabilities. Since they are voluntary, registration is optional.

Registries have been controversial because:

- Registries are time consuming and costly to update; they may be inaccurate if not regularly updated.
- Some people with disabilities do not want to register.
- Registries track home addresses; the registered person may not be home when disaster strikes.
- Confidentiality, privacy and misuse of registry information can be a concern.
- Registries do not include people who are temporarily in the community, for example for work, to attend school, or to do business.
- Registries likely do not include people with temporary disabilities, who may have access and functional needs for a specific period of time.
- People may feel that a registry implies a promise of evacuation, rescue, or priority, which is inaccurate, and they might not self-prepare.

- There is limited research regarding registry effectiveness and best practices.

While a registry might paint a partial picture of local disability demographics, it would be at best a floor, not a ceiling, and at worst highly inaccurate. A more comprehensive way to assess disability demographics and make outreach to the disability community is to use multiple means. For example, for counties and many smaller community structures, the American Community Survey (https://www.census.gov/programs-surveys/acs) has census based information regarding six categories of functional need (e.g., related to hearing, sight, difficulty concentrating, remembering things and making decisions, etc.). This would provide some demographic data. A parallel strategy would be to engage with local disability and other community based organizations for collaboration. These organizations know the constituents they serve and could act as liaisons for sharing information to and from their constituents. Of course, some people with disabilities do not receive human or social services and so it would be important to provide relevant information through the usual multiple and accessible public channels and in multiple accessible formats as well [68,69].

Portions of this text box come from Association of University Centers on Disabilities (AUCD). Prepared4ALL whole community inclusive emergency planning. Association of University Centers on Disabilities (AUCD); 2021 [cited 2021 Oct 21]. Available from: https:// nationalcenterdph.org/our-focus-areas/emergency-preparedness/prepared4all/online-training/. Used with permission.

Tribal emergency planning

In the sovereign nations which often cover large areas of land in the United States, tribal governments play an important role in emergency planning. According to the Centers for Disease Control and Prevention (CDC), as sovereign entities, federally recognized Native American and Alaska Native tribes have the authority to develop their own public health preparedness and response strategies and plans [70]. Where previously tribal governments had to go through the state in which they were located for a governor's request for federal emergency assistance, currently, under the federal Stafford Act, the tribal chief executive can directly request a presidential emergency declaration [70,71]. Under a presidential Executive Order, FEMA is mandated to regularly consult with recognized tribal governments which FEMA does as part of a policy of recognizing a nation to nation relationship with tribal governments [8,72,73].

Disability inclusion gaps in the incident command system and local emergency plans

The National Incident Management System (NIMS) doesn't mention people with disabilities. Within the Incident Command System (ICS) there is no set place to address the needs of disabled people and others with access and functional needs. For example, there are 15 Emergency Support Functions (ESF) [10] and no specific ESF includes responsibility for the disaster needs of this population. Since the command and control system for managing emergency incidents omits this responsibility, this means that in practice people with disabilities may be overlooked or that one ESF may assume that another part of the ICS will take responsibility for this population. As well, merely assigning disability-related issues to the public health and medical service responsibilities ESF (ESF 8), presupposes that disability is a medical issue, rather than taking a broader, more inclusive view of disability as a demographic. Best practices would assume that all ESFs would incorporate the needs of disabled people, as the ESFs should do for the general population as well. However, because of ongoing inequities it may be important for an emergency plan to specifically highlight that attention will be given to the disaster-related needs of the population with disabilities by making a clear statement about inclusion and identifying the entity with responsibility for ensuring that these needs are met.

The likely disaster needs of people with disabilities and others with access and functional needs are often overlooked in drafting emergency plans. When people with disabilities are absent from the emergency planning table, their needs will be excluded from emergency plans, and those plans may include such unrealistic pieces of advice such as to use a fire extinguisher and "pull, aim, squeeze, and sweep" rather than "stop, drop, and roll" to address fires, as described further below. Ensuring that the needs of everyone in the community, including people with disabilities and others with access and functional needs, are met by emergency plans is critical, and a matter of both equity and law.

Reviewing an emergency plan for disability inclusion gaps

Consider the following questions when reviewing an emergency or public health preparedness plan for disability inclusion gaps.

- What assumptions, if any, does the plan make?
  - Does the plan assume that *someone else* should address emergency-related disability and other access and functional needs?
  - Does the plan assume that all community members can talk, walk, run, quickly follow directions, drive (and have cars)? Does the plan assume that community members have the ability, understanding, financial and other resources to take care of themselves during a disaster? If so, the plan would likely exclude many people with disabilities.
  - Does the plan assume that all community members speak, read and write English and communicate in the same way?
- Does the plan address planning and response related to disabled people and others with access and functional needs?
  - Where does the plan place responsibility for addressing the needs of people who use powered medical devices (e.g., ventilators) and require a constant electricity source?
  - Which organization is responsible for replacing lost or destroyed critical medical equipment?
  - Which organization is responsible for making mass care shelters physically and programmatically accessible (required under the Americans with Disabilities Act)?
- How much community input into the local plan(s) has there been? Stakeholders included in emergency and public health preparedness planning should comprise a diverse group representing the residents, students, workers and others in the community, plus local businesses and community-based organizations, health care providers, schools, and faith-based groups, among others. Some communities have voluntary emergency planning or response organizations, and outreach should be made to them too.

Credit: *Association of University Centers on Disability (AUCD), National Center on Disabilities in Public Health, Prepared4ALL Initiative. Used with permission. Association of University Centers on Disabilities (AUCD). Prepared4ALL whole community inclusive emergency planning. Association of University Centers on Disabilities (AUCD); 2021 [cited 2021 Oct 21]. Available from: https://nationalcenterdph.org/our-focus-areas/emergency-preparedness/prepared4all/online-training/.*

Other stakeholders: Disability organizations, voluntary disaster organizations and emergency responders

Increasingly state and local disability organizations, voluntary disaster organizations, and volunteer responders are playing key roles related to emergencies and disasters. The importance of, and potential disaster roles for, community disability organizations is an important component of whole community emergency planning and response. Voluntary disaster organizations can also contribute to disability inclusion and are discussed below. Other potential allies and collaborators would be state and local ADA Coordinators. The ADA requires that state and local governments with 50 or more employees appoint an ADA Coordinator to oversee, implement and evaluate ADA compliance, including that related to emergency management and public health preparedness [74]. ADA Coordinators with knowledge and experience could play an important role in strengthening disability inclusion in local emergency management and public health preparedness, although at least one study has identified a number of gaps in this area. Researchers examined the role ADA Coordinators played in fostering accessibility in emergency management. Results of a survey of ADA Coordinators in four states and the Pacific Islands (FEMA Region IX) indicate that only a very small minority of ADA Coordinators (9%, $N = 131$) saw it as their responsibility to assure disability access in emergency management, only 22% ($N = 131$) felt knowledgeable about the topic, and only 36% ($N = 131$) believed they had received adequate training on disability access in emergency management [75].

Some states and communities have volunteer responders and/or emergency planning advisory groups. For example, Oregon's approach incorporates a Disability Emergency Management Advisory Council (DEMAC), which is active during emergencies, including during the COVID-19 pandemic. This group is a cross-disability statewide council, the focus of which is to bring disability community lived expertise to state emergency management and public health preparedness processes [76]. The Council is administered and funded collaboratively by three state agencies: the Oregon Department of Human Services (DHS), state Health Authority (OHA), and state Office of Emergency Management (OEM). Other states or communities have Local Emergency Planning Committees (LEPC). Originally related to Superfund environmental compliance and response to hazardous material spills, some of the LEPCs currently have broader scopes related to emergencies and disasters [77,78]. In addition to these groups, there are community-based voluntary planning and response organizations, such as Community Organizations Active in Disasters (COAD) [79] or Voluntary Organizations Active in Disasters (VOAD) [80], both comprised of community-based organizations. Medical Reserve Corps (MRC) [81] are generally medical and public health volunteer community member responders, and Community Emergency Response Teams (CERT) [82] or Citizens Corps Programs (CCP) [83] are other community volunteer response groups. There are rich opportunities for these organizations to make outreach to and include disabled individuals as members as well as provide response services to people with disabilities who are disaster survivors.

Structural challenges and silos (Fig. 6)

Silos within the system of systems

The emergency management silo includes FEMA at the US Department of Homeland Security, the 10 FEMA regions each made up of a number of states/territories, the state/territory emergency management agency, homeland security or emergency management regions within a state/territory, and the tribal, city, town, and/or county emergency management department.

The public health preparedness silo includes ASPR at the US Department of Health and Human Services (as well as other Health and Human Services offices such as the CDC, FDA, etc.), the 10 Health and Human Services regions, the state/territory public health department, ASPR regional Healthcare Coalitions within states, and tribal, city, town, and/or county departments of health.

FIG. 6

Silos among and between the US emergency management and public health preparedness systems.

Credit: Association of University Centers on Disability (AUCD), National Center on Disabilities in Public Health, Prepared4ALL Initiative. Used with permission. Association of University Centers on Disabilities (AUCD). Prepared4ALL whole community inclusive emergency planning. Association of University Centers on Disabilities (AUCD); 2021 [cited 2021 Oct 21]. Available from: https:// nationalcenterdph.org/our-focus-areas/emergency-preparedness/prepared4all/online-training/.

As might be expected with a system of systems, authority is widely distributed. FEMA or ASPR can control the regional systems that they created [23]. However, the state systems are independent of direct federal control [84], but indirect control may be exercised via grant funding. This means that FEMA and ASPR as the federal preparedness authorities, cannot issue mandatory regulations that states and localities must follow. The federal system can only provide guidance or make suggestions to state and local systems. At the local level, emergency management agencies are also generally independent of state emergency management agencies [84]. The situation with local public health departments is more complex and the level of control the state exercises over local public health varies. In three states, the local public health departments are governed jointly with the state public health department, while seven states have local public health departments that are units of the state public health department. In all other states, there is more than one type of governance for local health agencies, including local health departments that are independent of state control [85].

These relationships indicate that systems change to expand equity and disability inclusion on the local level need not necessarily rely on federal or state authority. FEMA and ASPR cannot mandate that state emergency management and public health preparedness become more inclusive. Likewise, without a state statute authorizing it, state emergency management agencies can't mandate that local emergency management become more inclusive. This holds true for a majority of state and local public health departments as well. This structure has also created silos leading to gaps and conflicts both vertically, across federal-state-local levels, and horizontally, between public health and emergency management agencies at every level. Consequently, other methods of interaction must be employed to reach these goals. Cross-sector collaboration, where entities come together to focus their knowledge and expertise on a common problem, offers one solution for this problem. Cross sector collaboration is considered a best practice among governmental agencies, including public health and emergency management [84,86,87] and is built into the National Preparedness Goal. This kind of collaboration represents a whole of government approach, although in practice this may be difficult to achieve [7].

The current organization of the emergency system with its multilayered and dually focused structure on public safety/emergency management and public health makes collaboration and communication between and among all levels of the systems a challenge. Overlapping roles and gaps in resources create sometimes troubled relationships between agencies both on the same level and between levels. The integration of resources and services dependent on the cooperation between agencies may suffer when agencies lack effective collaboration and communication. Best practices would suggest that enhanced coordination between public health and emergency management entities would lead to a more efficient and effective response and recovery in the event of a disaster. Collaboration during the planning phase before a disaster is particularly important. If collaborative relationships are not established prior to an event it may be impossible for entities to work together to respond to the event. Coordination challenges include the lack of interoperable communication systems as local communities often purchase their own communications systems and the different communication systems may not be able to link to each other. Practically, this means that the first responder radio communications in one community may not be able to connect with the radio communications of responders in other communities, making coordination during a disaster difficult. One role that many states have taken on is to work with local communities to coordinate communication systems to allow effective communications between communities [88,89].

As well, the dispersed authority of the US government makes a centralized, coordinated emergency response difficult. With the decentralization of powers, resources and decision making, all of the elements of emergency planning systems require cooperation and partnership to achieve the level of integration and coordination required to respond to the potential of a massive disaster. This is complicated by the lack of a standard for sharing resources between emergency planning agencies and regions affected by disasters. In turn this results in a lack of clarity about the federal disaster response to states and localities.

Traditional emergency plan development

Generally, emergency plans in the United States have been developed in a top down fashion, with local emergency managers creating an individual county, city or town's emergency plan without input from the public and the various stakeholder groups that should be involved. Traditional emergency response also involves a hierarchical, top-down paternalistic approach in which public administrators make decisions and in which the public is viewed as unable to help, likely to panic, and is often seen as hindering the work of the professionals [90–92]. Emergency plans are generally written by and for emergency managers rather than the public [90] and focus almost solely on those who can walk, run, talk, drive, and quickly follow directions [93]. This practice excludes many individuals with disabilities and other potentially vulnerable populations, leaving them without emergency plans that address their needs. As well, planners may make assumptions about residents' financial or other resources, but those assumptions may not be realistic. For example, just because a resident has a car and can evacuate, doesn't mean that they have the financial resources to support themselves during and after evacuation.

Historically, the United States has neglected the disaster needs of potentially vulnerable populations such as disabled people, children, people with low incomes, and racial and ethnic minorities [94] by excluding them from local emergency planning and omitting their likely emergency needs from local plans [95]. Furthermore, not meeting disability-related needs, for example by not providing accessible emergency shelters or disability accommodations within shelters, sign language interpreters, accessible emergency warnings and alerts, and other necessary accommodations during emergency response and recovery efforts, is a civil rights violation. Quigley [96] describes the experiences of individuals with disabilities during Hurricane Katrina (2005). In addition to that fact that 27% of New Orleans' population lacked a means of personal transportation to evacuate, over 20% of people with physical disabilities were unable to evacuate. Unfortunately, emergency planning did not incorporate the needs of those individuals.

One example of the kind of outdated emergency advice provided for people in wheelchairs instructs them to consider mounting a small personal use fire extinguisher on their wheelchairs where they could access it in the event of a fire. The advice directs the person to learn how to use it with the expression "pull, aim, squeeze, and sweep" rather than the "stop, drop, and roll" [97, p. 30] taught to the general public. While following such advice might be possible for some, it is inaccessible to others in addition to being impractical. In addition, the guidance does not help people with disabilities think through the feasibility of how to approach fire safety given their specific individual needs, strengths, environment, and support networks.

In 2005, during and after Hurricane Katrina in Louisiana, many voices brought the gaps in emergency planning and response for and with individuals with disabilities to the forefront. The National Council on Disability, an independent federal agency, increased the awareness of this issue when it published its seminal report, *Saving Lives: Including People with Disabilities in Emergency Planning* [97]. The report findings identify a number of areas of weakness in how emergency planning and response entities address the needs of people with disabilities. In particular, the report notes that lessons learned in previous disasters related to people with disabilities were not used to inform or change response in subsequent disasters [97]. Individuals with disabilities are experts based on their lived experience. Excluding people with disabilities from the planning and post disaster analysis risks losing the valuable information that they can provide about how best to meet their likely emergency needs, as well as losing their other strengths and skills. Inclusive emergency planning or "whole community" planning leverages the strengths and skills of people with disabilities in addition to other community members, non-profit organizations, and businesses to facilitate a smoother response and recovery.

The key findings of the report also note that planning and response information related to the needs of people with disabilities in an emergency should be integrated into general emergency management courses as well as providing more specific information tailored to people with disabilities [97]. Additionally, it identifies that emergency management and public health preparedness do not make use of the strengths and skills of the community based organizations that work with people with disabilities. Integrating disabled individuals and disability organizations into the planning process and fostering partnerships with local governments will strengthen emergency plans and help emergency response better meet the different needs of the disabled population. Stronger outreach, targeted technical assistance, and training initiatives focused on the population and their needs in an emergency are also recommended [97].

Disappointingly, in their extensive report covering disasters from 2017 and 2018 Roth, Kailes, and Marshall note some of the same civil rights violations, lack of equity, access and accommodations that existed in 2005 [98]. Their report outlines a continued lack of inclusive emergency planning for people with disabilities, including a lack of universal design (aimed at providing access to almost everyone) and the ongoing need for systems change. During disasters people with disabilities still encounter physical, programmatic, and communication barriers in locations used for disaster operations including shelters and recovery centers, first aid stations, mass feeding areas, and portable toilets. They also encounter barriers to evacuation and filing claims for disaster aid.

Roth, Kailes, & Marshall [98] offer the following story reported by the National Council on Independent Living (a disability self-advocacy umbrella organization) about lack of access, as one of many in their report. During a mandatory evacuation order in Texas a male wheelchair user stayed at a Red Cross Shelter starting on August 30.

> This shelter provided daily trips to Walmart and the laundromat, but the bus for these trips was not accessible to him or the other wheelchair users in the shelter, making these services completely inaccessible to them. On 6 September, the mandatory evacuation was lifted for the man's hometown, but the buses provided were not accessible. Two other wheelchair users were told to load their luggage on the inaccessible buses and that they would be transported by ambulance the next day, separately from their belongings. The man and two other wheelchair users were transported back home on 11 September, five days after the mandatory evacuation was lifted [98, p. 128].

Some progress in this realm has been made since Hurricane Katrina. On the federal level, after Hurricane Katrina advocacy efforts and media attention led then President George W. Bush to issue an Executive Order establishing a federal Interagency Coordinating Council on Emergency Preparedness and Individuals with Disabilities [99]. The Post-Katrina Emergency Management Reform Act of 2006 [37] also added a disability coordinator position to FEMA to begin to genuinely consider the needs of the disability population and include them in emergency management. While the same kind of advice that fails to consider the individual needs of people with disabilities can still be found today, updated more realistic guidance is also available, such as that mentioned above related to Massachusetts, and the functional needs self-preparedness advice from South Carolina. The disproportionate impact of COVID-19 on people with disabilities, with concomitant equity issues, has illustrated ongoing challenges but has also increased advocacy and the search for solutions.

Conclusion

This chapter provided a broad overview of the US "system of systems" for emergency management and public health preparedness, reviewing the guiding principles, structures, and roles and responsibilities at the federal, state, and local levels across public health preparedness and emergency management systems. Within this structure, there is no set place to address issues related to people with disabilities and others with access and functional needs, although some jurisdictions have included these populations within different parts of the ICS structure, in the Emergency Operations Center, and in other ways. The traditional process for developing local emergency plans consisted of one or a few professionals drafting the plan without much, if any, stakeholder input, a gap that leads to the exclusion of people with disabilities and others with access and functional needs and the creation of unrealistic guidance. The advice for people who use wheelchairs to "pull, aim, squeeze, and sweep" in a fire is an example of guidance that is developed without input by stakeholder with lived experience and knowledge of the feasibility of implementing the recommendations.

Localities, states/territories, and the federal government are increasingly planning more inclusively and with particular attention to the varied needs of the population, including people with disabilities. The example of the Oregon DEMAC advisory committee was given in this chapter. Chapter 10 about promising practices explores these improvements further. As emergency planning more consistently involves people with disabilities and allies, individual guidance and emergency plans should continue to improve and better account for the range of capabilities that people with disabilities have. As individuals with disabilities and allies continue to advocate for inclusive emergency planning and better educate themselves about self-preparedness, systems will continue to improve, people with disabilities will better choose an approach to personal planning that meets their individual needs, and systems can better plan to be able to meet the needs of diverse populations.

Questions for thought

(1) The American emergency management system is challenged by silos between emergency management and public health preparedness. Silos are common within and between organizations. One successful strategy to break down silos is to identify common department or organization missions. What are other strategies to break down silos?

(2) Vielot and Horney [100] report a local emergency preparedness coordinator position shared between a local emergency management department and a local public health department. What do you see as the benefits and challenges in having such a position?

(3) The concept of federalism, as presented in the US Constitution, undergirds the emergency management and public health preparedness planning structure in the United States. What do you think are the advantages and disadvantages of federalism related to emergency management and public health preparedness?

References

[1] Benjamin Franklin Historical Society. Union fire company. Benjamin Franklin Historical Society; 2014. [cited 2021 Dec 28]. Available from: http://www.benjamin-franklin-history.org/union-fire-company/.

[2] American Red Cross. American Red Cross founder Clara Barton. American Red Cross. [cited 2021 December 28]. Available from: https://www.redcross.org/content/dam/redcross/enterprise-assets/about-us/history/history-clara-barton-v5.pdf.

[3] Haddow GD, Bullock JA, Coppola DP. Introduction to emergency management. Cambridge, MA: Butterworth-Heinemann; 2021.

[4] United States Department of Homeland Security/FEMA. Emergency management authorities review. United States Department of Homeland Security/FEMA; 2018. [cited 2021 Dec 1]. Available from: https://emilms.fema.gov/IS0230d/FEM0101170text.htm.

[5] American Public Health Association. Our history. American Public Health Association; 2021. [cited 2022 Feb 1]. Available from: https://www.apha.org/about-apha/our-history.

[6] McKinney S, Papke ME. Public health emergency preparedness. Burlington, MA: Jones & Bartlett; 2019.

[7] United States Department of Homeland Security. National preparedness goal. Washington, DC: United States Department of Homeland Security; September, 2015. Available from: https://www.fema.gov/sites/default/files/2020-06/national_preparedness_goal_2nd_edition.pdf.

[8] United States Department of Homeland Security. National response framework. 4th ed. Washington, DC: United States Department of Homeland Security/FEMA; October 28 2019. Available from: https://www.fema.gov/sites/default/files/2020-04/NRF_FINALApproved_2011028.pdf.

[9] Landesman LY, Gershon RR, Gebbie EN, Merdjanoff AA. Landesman's public health management of disasters: The practice guide. 5th ed. Washington, DC: APHA Press; 2021.

[10] United States Department of Homeland Security FEMA. National incident management system. 3rd ed. United States Department of Homeland Security/FEMA; 2017. Available from: https://www.fema.gov/sites/default/files/2020-07/fema_nims_doctrine-2017.pdf.

[11] Katz R, Banaski JA. Essentials of public health preparedness and emergency management. 2nd ed. Burlington, MA: Jones & Bartlett; 2019.

[12] Ransom MM, Valladares LM, editors. Public health law concepts and cases. New York, NY: Springer Publishing; 2022.

[13] The Annenberg Public Policy Center of the University of Pennsylvania. Federalism. Annenberg Classroom; 2021. [cited 2021 Dec 28]. Available from: https://www.annenbergclassroom.org/glossary_term/federalism/.

[14] Hunter ND. The law of emergencies: Public health and disaster management. Butterworth-Heinemann; 2018.

[15] U.S. Constitution. Tenth Amendment.

[16] Congressional Research Service. Congressional primer on responding to and recovering from major disasters and emergencies, June 3 2020. [cited 2021 Dec 29]. Available from: https://sgp.fas.org/crs/homesec/R41981.pdf.

[17] Lindsay B, Congressional Research Service. Federal emergency management: A Brief introduction, 2012 Nov. Available from: https://sgp.fas.org/crs/homesec/R42845.pdf.

[18] Flanagan BE, Gregory EW, Hallisey EJ, Heitgerd JL, Lewis B. A social vulnerability index for disaster management. J Homel Secur Emerg Manag 2011;8(1):1–22.

[19] Bexar County, Texas Emergency Management. The five phases of emergency management. Bexar County Emergency Management; 2021. [cited 2021 Dec 28]. Available from: https://www.bexar.org/694/Five-Phases.

[20] Congressional Research Service. A brief overview of FEMA's public assistance program, March 8 2021. Available from: https://crsreports.congress.gov/product/pdf/IF/IF11529.

[21] United States Department of Homeland Security. What is FEMA public assistance? 2019. Updated 2021. Available from: https://www.fema.gov/press-release/20210318/what-fema-public-assistance.

[22] United States Department of Homeland Security FEMA. Individuals and households program, 2021. Available from: https://www.fema.gov/assistance/individual/program.

[23] Wolf-Fordham S. Integrating government silos: local emergency management and public health department collaboration for emergency planning and response. Am Rev Public Adm 2020;50(6–7):560–7.

[24] United States Department of Homeland Security FEMA. Locally executed, state managed, federally supported recovery, 2021. Available from: https://www.fema.gov/case-study/locally-executed-state-managed-federally-supported-recovery.

[25] FEMA. Emergency Management Institute (EMI). Overview. FEMA Emergency Management Institute; 2019. [cited 2021 Dec 26]. Available from: https://training.fema.gov/history.aspx.

[26] United States Department of Health and Human Services Office of the Assistant Secretary for Preparedness and Response. About ASPR. Public Health Emergency; 2021. Available from: https://aspr.hhs.gov/AboutASPR/Pages/default.aspx.

[27] The Town of Phillipsburg, New Jersey. Emergency management history. Phillipsburg, NJ: The Town of Phillipsburg; 2019. Available from: http://www.phillipsburgnj.org/wp-content/uploads/2015/01/ACF4A8.pdf.

[28] United States Department of Homeland Security FEMA. In: About us. FEMA; 2021. [cited 2021 Dec 15]. Available from: https://www.fema.gov/about.

[29] United States Department of Homeland Security FEMA. 2018-2022 FEMA strategic plan, 2018. Available from: https://www.fema.gov/sites/default/files/2020-03/fema-strategic-plan_2018-2022.pdf.

[30] FEMA. 2022-2026 FEMA strategic plan, 2021. Available from: https://www.fema.gov/sites/default/files/documents/fema_2022-2026-strategic-plan.pdf.

[31] FEMA. Disaster Recovery Reform Act of 2018. United States Department of Homeland Security FEMA; 2021. Available from: https://www.fema.gov/disaster/disaster-recovery-reform-act-2018.

[32] United States Department of Health and Human Services. Pandemic and All Hazards Preparedness Act, 2014. Available from: https://www.phe.gov/Preparedness/legal/pahpa/Pages/default.aspx.

[33] United States Department of Health and Human Services Office of the Assistant Secretary for Preparedness and Response. ASPR mission and vision. Public Health Emergency; 2021. Available from: https://www.phe.gov/about/aspr/Pages/default-archive.aspx.

[34] United States Department of Health and Human Services Office of the Assistant Secretary for Preparedness and Response. ASPR highlights. United States Department of Health and Human Services Office of the Assistant Secretary for Preparedness and Response; 2021. [cited 2021 Dec 20]. Available from: https://www.phe.gov/about/Pages/highlights.aspx.

[35] United States Department of Health and Human Services Office of the Assistant Secretary for Preparedness and Response. 2017-2022 Health care preparedness and response capabilities, 2016. Available from: https://www.phe.gov/preparedness/planning/hpp/reports/documents/2017-2022-healthcare-pr-capablities.pdf.

[36] United States Department of Health and Human Services Office of the Assistant Secretary for Preparedness and Response. General overview of healthcare coalitions, 2022. Available from: https://files.asprtracie.hhs.gov/documents/aspr-tracie-general-overview-hccs.pdf.

[37] Post-Katrina Emergency Management Reform Act of 2006., 2006. Public Law 109–295 Oct 4, 2006. Available from: https://www.govinfo.gov/content/pkg/PLAW-109publ295/pdf/PLAW-109publ295.pdf.

[38] United States Department of Homeland Security FEMA. Office of disability integration and coordination, 2022. Retrieved from: https://www.fema.gov/about/offices/disability.

[39] United States Department of Homeland Security FEMA. Office of Equal Rights. United States Department of Homeland Security FEMA; 2021. [cited 2021 Nov 20]. Available from: https://www.fema.gov/about/offices/equal-rights.

[40] United States Department of Health and Human Services Office of the Assistant Secretary for Preparedness and Response. At-risk individuals, behavioral health & community resilience (ABC). United States Department of Health and Human Services Office of the Assistant Secretary for Preparedness and Response; 2021. [cited 2021 Dec 20]. Available from: https://www.phe.gov/Preparedness/planning/abc/Pages/default.aspx.

[41] United States Department of Health and Human Services Office for Civil Rights. About us. United States Department of Health and Human Services Office for Civil Rights; 2021. [cited 2021 Dec 26]. Available from: https://www.hhs.gov/ocr/about-us/index.html.

[42] United States Department of Health and Human Services Office for Civil Rights. OCR resolves three discrimination complaints after medstar health system ensures patients with disabilities can have support persons in health care settings during Covid-19 pandemic. United States Department of Health and Human Services Office for Civil Rights; 2021. [cited 2021 Dec 26]. Available from: https://www.hhs.gov/about/news/2021/02/25/ocr-resolves-three-discrimination-complaints-after-medstar-health-system.html.

[43] Centers for Disease Control and Prevention. Center for preparedness and response. Centers for Disease Control and Prevention; 2021. [cited 2021 Dec 20]. Available from: https://www.cdc.gov/cpr/index.htm.

[44] Centers for Disease Control and Prevention. Public health emergency preparedness and response capabilities, October, 2018. Updated January 2019. Available from: https://www.cdc.gov/cpr/readiness/00_docs/CDC_PreparednesResponseCapabilities_October2018_Final_508.pdf.

[45] Centers for Disease Control and Prevention. Access and functional needs toolkit: Integrating a community partner network to inform risk communication strategies. Centers for Disease Control and Prevention; 2021. [cited 2021 Dec 20]. Available from: https://www.cdc.gov/cpr/readiness/afntoolkit.htm.

[46] Centers for Disease Control and Prevention. Disability and health emergency preparedness. Centers for Disease Control and Prevention; 2020. [cited 2021 Dec 20]. Available from: https://www.cdc.gov/ncbddd/disabilityandhealth/emergencypreparedness.html.

[47] United States Department of Homeland Security FEMA. Developing and maintaining emergency operations plans comprehensive preparedness guide (CPG). 3rd ed; 2021. [Internet]; Available from: https://www.fema.gov/sites/default/files/documents/fema_cpg-101-v3-developing-maintaining-eops.pdf.

[48] Centers for Disease Control and Prevention. CDC's public health emergency preparedness program: Every response is local. Centers for Disease Control and Prevention; October 7 2021. [cited 2021 Dec 20]. Available from: https://www.cdc.gov/cpr/whatwedo/phep.htm.

[49] Centers for Disease Control and Prevention. On-Trac. Centers for Disease Control and Prevention; 2020. [cited 2021 Dec 20]. Available from: https://www.cdc.gov/cpr/readiness/on-trac.htm.

[50] Executive Office of Public Safety and Security (EOPSS)., 2021. [cited 2021 Dec 20]. Available from: https://www.mass.gov/orgs/executive-office-of-public-safety-and-security.

[51] The South Carolina Office of The Adjutant General. The Military Department of South Carolina State Operations. The South Carolina Office of The Adjutant General; 2017. [cited 2021 Dec 20]. Available from: https://sctag.org/.

[52] South Carolina Emergency Management Division. Citizens with functional needs. South Carolina Emergency Management Division; 2021. [cited 2021 Dec 20]. Available from: https://www.scemd.org/prepare/your-emergency-plan/family-disaster-plan/citizens-with-functional-needs/.

[53] Massachusetts Emergency Management Agency. Emergency planning for people with disabilities and others with access and functional needs a resource guide for local emergency management directors/agencies in Massachusetts, 2019. Available from: https://www.mass.gov/files/documents/2019/08/12/2019%20MEMA%20AFN%20Resource%20Guide.pdf.

[54] Anon. Office of Preparedness and Emergency Management (OPEM)., 2021. [cited 2021 Dec 21]. Available from: https://www.mass.gov/orgs/office-of-preparedness-and-emergency-management.

[55] Commonwealth of Massachusetts Department of Public Health. Show me booklet—A communication tool for emergency shelters. Massachusetts Department of Public Health; 2021. [cited 2021 Dec 21]. Available from: https://massclearinghouse.ehs.state.ma.us/PROG-EMGPREP/MS2322.html.

[56] Commonwealth of Massachusetts Department of Public Health. Emergency planning tool stay aware. Be prepared, 2021. [cited 2021 Dec 21]. Available from: https://www.mass.gov/lists/stay-aware-be-prepared-emergency-preparedness-materials#booklet-.

[57] Massachusetts Department of Public Health. Emergency preparedness populations planning tool, 2022. [cited 2022 Jan 5]. Available from: https://dphanalytics.hhs.mass.gov/ibmcognos/bi/?perspective=authoring&pathRef=.public_folders%2FMEPHTN%2Fdph%2FOPEM%2BReport%2FEmergency%2BPreparedness%2BPlanning%2BTool&id=iC13BD38003BD46A39585927F08FCF943&closeWindowOnLastView=true&ui_appbar=false&ui_navbar=false&objRef=iC13BD38003BD46A39585927F08FCF943&action=run&format=HTML&cmPropStr=%7B%22id%22%3A%22iC13BD38003BD46A39585927F08FCF943%22%2C%22type%22%3A%22report%22%2C%22defaultName%22%3A%22Emergency%20Preparedness%20Planning%20Tool%22%2C%22permissions%22%3A%5B%22execute%22%2C%22read%22%2C%22traverse%22%5D%7D.

[58] S. C. Department of Health and Environmental Control (DHEC). Disaster preparedness. S.C. Department of Health and Environmental Control (DHEC); 2019. [cited 2021 Dec 21]. Available from: https://scdhec.gov/disaster-preparedness.

[59] S. C. Department of Health and Environmental Control (DHEC). Individuals with functional needs—How to prepare for emergency. S.C. Department of Health and Environmental Control (DHEC); 2019. [cited 2021 Dec 30]. Available from: https://scdhec.gov/disaster-preparedness/individuals-functional-needs-how-prepare-emergency.

[60] South Carolina Department of Health and Environmental Control. Special medical needs shelters a preparedness guide, 2018 Aug. Available from: https://scdhec.gov/sites/default/files/Library/ML-025588.pdf.

[61] FEMA. Guidance on planning for integration of functional needs support services in general population shelters, 2010 Nov. Available from: https://www.fema.gov/pdf/about/odic/fnss_guidance.pdf.

[62] Town of Bedford, New Hampshire. Community. Town of Bedford, New Hampshire; 2021. [cited 2021 Dec 20]. Available from: https://www.bedfordnh.org/31/Community.

[63] Town of Bedford, New Hampshire. Emergency management. Town of Bedford, New Hampshire; 2021. [cited 2021 Dec 30]. Available from: https://www.bedfordnh.org/174/Emergency-Management.

[64] PEMA. County and local emergency operations plans. PEMA; 2021. [cited 2021 Dec 30]. Available from: https://www.pema.pa.gov/Preparedness/Planning/Community-Planning/Pages/County-Local-EOP.aspx.

[65] U.S. Census Bureau. American Community Survey 1-year estimates, 2019. Census Reporter Profile page for Hennepin County, MN. [cited 2021 Dec 31]. Available from: http://censusreporter.org/profiles/05000US27053-hennepin-county-mn/.

[66] Hennepin County Minnesota. Emergency management. Hennepin County, Minnesota; 2021. [cited 2021 Dec 31]. Available from: https://www.hennepin.us/residents/emergencies/emergency-management.

[67] State of Montana. Montana emergency response framework, 2017. Available from: https://des.mt.gov/Preparedness/MERF-Version-3_1_Final.pdf.

[68] Association of University Centers on Disabilities (AUCD). Prepared4ALL whole community inclusive emergency planning. Association of University Centers on Disabilities (AUCD); 2021. [cited 2021 Oct 21]. Available from: https://nationalcenterdph.org/our-focus-areas/emergency-preparedness/prepared4all/online-training/.

[69] Kailes JI, Enders A. Emergency registries for people with access and functional needs. June Isaacson Kailes, Disability Policy Consultant; 2014. [cited 2021 Oct 21]. Available from: http://www.jik.com/d-rgt.html.

[70] Centers for Disease Control and Prevention. Tribal emergency preparedness law. Centers for Disease Control and Prevention; 2017. [cited 2021 Dec 30]. Available from: https://www.cdc.gov/phlp/publications/topic/briefs/brief-tribalemergency.html.

[71] Anon. Sandy Recovery Improvement Act of 2013., 2013. 42 U.S.C. §§ 5170(b)(1), 5191(c)(1). Available from: https://www.fema.gov/disaster/sandy-recovery-improvement-act-2013.

[72] United Stated Department of Homeland Security FEMA. Tribal affairs. US Department of Homeland Security FEMA; 2021. [cited 2021 Dec 20]. Available from: https://www.fema.gov/about/organization/tribes.

[73] United States Department of Homeland Security FEMA. FEMA tribal consultation policy [Internet]. United States Department of Homeland Security FEMA; 2019. [cited 2021 Dec 20]. Available from: https://www.fema.gov/sites/default/files/2020-04/CLEAN_FP_101-002-2_Tribal_Policy_June_2019_Signed.pdf.

[74] Great Plains ADA Center. ADA coordinators [Internet]. Great Plains ADA Center University of Missouri-Architectural Studies Department; 2021. [cited 2021 Dec 20]. Available from: https://www.gpadacenter.org/ada-coordinators.

[75] Gershon R, Kraus L, Merdjanoff A, Nwankwo E, Zhi Q. ADA coordinators role in jurisdictional emergency management in region 9: Who is responsible? 2021 Apr 14. Available from: https://depts.washington.edu/adakt/sos/index.php?day=2&tz=PDT.

[76] Oregon Department of Human Services. Disability Emergency Management Advisory Committee (DEMAC) [Internet]. Oregon Department of Human Services; 2021. [cited 2021 Dec 20]. Available from: https://www.oregon.gov/dhs/EmergencyManagement/Pages/DEMAC.aspx.

[77] Chatham County. About LEPC. Chatham County local emergency planning committee; 2021. [cited 2021 Dec 20]. Available from: https://lepc.com/about-lepc/.

[78] Tuscaloosa County Emergency Management. LEPC semi-annual meeting notice. Tuscaloosa County Emergency Management; 2021. [cited 2021 Dec 20]. Available from: https://www.tuscaloosacountyema.org/partners/lepc.

[79] City of Portland, Oregon. Community organizations active in disaster (COAD), 2018. [cited 2021 Dec 20]. Available from: https://www.portland.gov/pbem/coad.

[80] National Voluntary Organizations Active in Disasters. National voluntary organizations active in disasters [Internet]. National Voluntary Organizations Active in Disasters; 2020. [cited 2021 Dec 21]. Available from: https://www.nvoad.org/.

[81] U.S. Department of Health and Human Services Office of the Assistant Secretary for Preparedness and Response. The medical reserve corps. U.S. Department of Health and Human Services Office of the Assistant Secretary for Preparedness and Response; 2021. [cited 2021 Dec 21]. Available from: https://www.phe.gov/mrc/Pages/default.aspx.

[82] ReadyGov. Community emergency response team, 2011. [cited 2021 Dec 20]. Available from: https://www.ready.gov/cert.

[83] ReadyGov. Citizen corps, 2021. [cited 2021 Dec 20]. Available from: https://www.ready.gov/citizen-corps.

[84] Kapucu N. Emergency management: whole community approach. In: Encyclopedia of public administration and public policy. 3rd ed. Boca Raton, FL: CRC Press Taylor & Francis Group; 2015. Available from: https://www.researchgate.net/publication/271500745_Emergency_Management_Whole_CommunityApproach.

[85] National Association of City & County Health Officials. National profile of local health departments, 2019. Available from: https://www.naccho.org/uploads/downloadable-resources/Programs/Public-Health-Infrastructure/NACCHO_2019_Profile_final.pdf.

[86] Howell P. A whole-of-government approach: embedding disaster resilience into municipal operations. PM Magazine 2020 May 1. Available from: https://icma.org/articles/pm-magazine/whole-government-approach-embedding-disaster-resilience-municipaloperations.

[87] Warm D. Local government collaboration for a new decade: risk, trust, and effectiveness. State Local Gov Rev 2011;43(1):60–5. https://doi.org/10.1177/0160323X1140043.

[88] U.S. Department of Homeland Security FEMA. Wireless communications interoperability awareness guide. Available from: https://www.cisa.gov/sites/default/files/publications/Wireless_Communications_Interoperability_Awareness_Guide.pdf.

[89] Cybersecurity and Infrastructure Security Agency. Emergency communications. Cybersecurity and Infrastructure Security Agency; 2022. [cited 2022 Jan 2]. Available from: https://www.cisa.gov/emergency-communications.

[90] Kaufman R, Bach R, Riquelme J. Engaging the whole community in the United States. In: Bach R, editor. Crismart volume I strategies for supporting community resilience: Multinational experiences. Stockholm, Sweden: Swedish Defence University; 2015. p. 151–86. Available from: https://www.preventionweb.net/publications/view/43699.

[91] Schoch-Spana M, Chamberlain A, Franco C, Gross J, Lam C, Mulcahy A, Nuzzo JB, Toner E, Usenza C. Disease, disaster, and democracy: the public's stake in health emergency planning. Biosecur Bioterror 2006;4(3):313–9. Available from: https://www.centerforhealthsecurity.org/our-work/events-archive/2006_disease_disaster_democracy/Conference%20Report.pdf.

[92] Schoch-Spana M, Franco C, Nuzzo JB, Usenza C. Community engagement: leadership tool for catastrophic health events. Biosecur Bioterror 2007;5(1):8–25.

[93] Kailes J, Enders A. Moving beyond "special needs": a function-based framework for emergency management and planning. J Disabil Policy Stud 2007;17(4):230–7. https://doi.org/10.1177/10442073070170040601.

[94] Horney J, Nguyen M, Salvesen D, Tomasco O, Berke P. Engaging the public in planning for disaster recovery. Int J Disaster Risk Reduct 2016;17:33–7.

[95] Sherry N, Harkins AM. Leveling the emergency preparedness playing field. J Emerg Manag 2011;9(6):11–5.

[96] Quigley W. Thirteen ways of looking at Katrina: human and civil rights left behind again. Tulane Law Rev 2007;81(4):955–1017.

[97] National Council on Disability. Saving lives: Including people with disabilities in emergency planning, 2005. Available from: https://ncd.gov/sites/default/files/Documents/saving_lives.pdf.

[98] Roth M, Kailes JI, Marshall M. Getting it wrong: An indictment with a blueprint for getting it right disability rights, obligations and responsibilities before, during and after disasters (edition 1), 2018 May. Available from: http://www.disasterstrategies.org/index.php/news/partnership-releases-2017-2018-after-action-report.

[99] United States Department of Homeland Security. Interagency Coordinating Council on Emergency Preparedness and Individuals with Disabilities Executive Order 13347 An update on activities and achievements. October 2006-July 2008, 2010. Available from: https://www.dhs.gov/sites/default/files/publications/crcl-icc-report-october06-july08.pdf.

[100] Vielot NA, Horney JA. Can merging the roles of public health preparedness and emergency management increase the efficiency and effectiveness of emergency planning and response? Int J Environ Res Public Health 2014;11(3):2911–21.

CHAPTER

Legal issues related to emergencies and disasters: Antidiscrimination and other selected issues

7

Katherine Snyder[a] and Susan Wolf-Fordham[b,c]

[a]*Independent Contributor,* [b]*Consultant, Association of University Centers on Disabilities, Silver Spring, MD, United States,* [c]*Adjunct Faculty of Public Health, Massachusetts College of Pharmacy and Health Sciences, Boston, MA, United States*

Source: Pexels

Integrating Mental Health and Disability Into Public Health Disaster Preparedness and Response
https://doi.org/10.1016/B978-0-12-814009-3.00003-9

Learning objectives

- Summarize the key features of the Americans with Disabilities Act (ADA), Section 504 of the Rehabilitation Act of 1973 (Section 504) and other civil rights laws that apply in emergencies, disasters and pandemics
- Describe examples of: "reasonable modifications," "most integrated setting," "fundamental alteration" of services, programs, or activities and "undue financial and administrative burdens" under the law
- Explain lessons learned from court decisions, legal settlements, and US Department of Justice compliance investigations related to disability and emergency management or public health preparedness

Introduction

The law is an important tool to strengthen emergency management and public health disaster preparedness and response [1], support equity, and policies and practices that promote individual and community resilience. The US Constitution established the American government structure with its separation of powers and principle of federalism. The federal government only has those powers specified in the Constitution, and any federal powers the Constitution does not enumerate are reserved to the states [2]. This is the source of the states' responsibility and authority to protect, preserve and promote the health and welfare of state residents via, among other areas, public health and emergency management. This authority is known as a state's "police power" [3]. States can and do delegate their "police power" authority to local governments and this is the legal basis for local emergency management and public health [3]. Unlike the discussion of the US emergency management system in Chapter 6, this chapter doesn't focus on legal structures. Rather, the chapter examines law as a tool to effect equity in emergency management, public health preparedness, and disaster service delivery, and provides an overview of key antidiscrimination statutes, court decisions, and settlements to illustrate the issues and identify where opportunities for improvement lie.

The chapter begins with a description of disability antidiscrimination legal principles in order to provide context for the rest of the chapter. What follows is a brief description of the key federal disability antidiscrimination statutes, and the requirements of "physical access," "program access," and "effective communication" for people with disabilities [4–6]. The discussion moves to a description of key class action lawsuits (a lawsuit brought by one or a few people on behalf of a larger group of people with the same legal interest) related to disability discrimination in emergency management, emergency plans, and the planning process. The chapter next describes a US Department of Justice Americans with Disabilities Act (ADA) compliance review process that focuses on local emergency management adherence to the ADA, among other issues. The class action and compliance review settlements offer potential solutions to close disability inclusion-related gaps in emergency management, as do two other court cases related to "effective communication." During pandemics or other times when medical resources may be scarce, healthcare institutions may need to make decisions about allocating these resources. The chapter briefly describes allocating scarce medical resources in relation to disabled people. Because the issue of intersectionality (i.e., that someone identifies as being a member of more than one historically marginalized group) is critical to understanding disability discrimination, the chapter looks at antidiscrimination laws related to populations other than those with disabilities. This chapter ends with a discussion of whether laws can be waived during emergencies and disasters, using the Health Insurance Portability and Accountability Act (HIPAA) privacy and health information security law and antidiscrimination laws as examples.

An important lesson from this chapter is that there may be more than one way to provide equal access and more than one reasonable modification needed to address a particular situation. Equal access is discussed below relating to effective communication, auxiliary aids and services. Different kinds of reasonable modifications can be illustrated in the following example. Imagine a community emergency planning meeting. Some of the reasonable modifications (adaptations or accommodations) for the meeting might include: (1) moving around furniture in the meeting room so there would be space for people using wheelchairs; (2) providing an agenda in a simplified text format with explanatory images to make for easier reading by someone with limited English literacy; (3) providing meeting notes in English plus another language; and (4) previewing the agenda with someone with an intellectual disability to better enable their fuller participation in the meeting. An important corollary to these concepts is that two people with the same diagnosis or what appear to be the same needs may in fact require different reasonable modifications. For example, some people with an autism diagnosis may need seating apart from other people at the community meeting, while others may need to participate remotely, and still others may need no modifications at all.

Overview of key nondiscrimination concepts

While the nondiscrimination concepts in this chapter might seem complex, the following list provides a good overview [7] (Fig. 1).

For emergency managers and public health preparedness planners it is important to remember that there is often no one right reasonable modification to ensure equity and inclusion. This is one reason why advance planning, using the CMIST (communication—maintaining health—independence—support, safety, self-determination—transportation) framework to address access and functional needs issues, and actively including disabled people in all parts of the emergency management and public health preparedness life cycle is so important. Making reasonable modification, effective communication, program or physical access decisions on the spur of the moment, in the middle of a disaster, may be inefficient at best and less likely to be as effective as more thoughtfully made decisions. But it is also important to remember that surprises will always occur during emergencies and disasters. Some disaster consequences are unforeseen and so there will always be unexpected events. A goal of advance planning is to minimize the likely number of surprises during emergency response and recovery.

Disability definition and antidiscrimination standards
Which law applies?

The main disability antidiscrimination laws this chapter discusses include: (1) The Americans with Disabilities Act of 1990 (ADA), particularly Title II which applies to state and local (city, town, county) emergency programs, services and activities including emergency management and public health preparedness [4]; (2) the ADA Amendments Act of 2008 (ADAA), which amended the ADA [5]; and (3) Section 504 of the Rehabilitation Act of 1973 (Section 504) [6]. Emergency managers and public health preparedness planners will see these three laws regularly mentioned in federal, state and local guidance

1. "Self-Determination – People with disabilities are the most knowledgeable about their own needs." Their decisions must be honored.

2. "No 'One-Size-Fits-All' – People with disabilities do not all require the same assistance and do not all have the same needs...."

3. "Equal Opportunity – People with disabilities must have the same opportunities to benefit from emergency programs, services, and activities as people without disabilities." There must be "equivalent choices" for disabled people.

4. "Inclusion – People with disabilities have the right to participate in and receive the benefits of emergency programs, services, and activities provided by governments, private businesses, and nonprofit organizations...."

5. "Integration – Emergency programs, services, and activities typically must be provided in an integrated setting...." This means, for example, that absent acute medical need disabled people should be housed in the same emergency shelters as everyone else.

6. "Physical Access – Emergency programs, services, and activities must be provided at locations that all people can access, including people with disabilities."

7. Program Access – "People with disabilities must be able to access and benefit from emergency programs, services, and activities equal to the general population."

8. "Effective Communication – People with disabilities must be given information that is comparable in content and detail to that given to the general public. It must also be accessible, understandable and timely."

9. "Program Modifications – People with disabilities must have equal access to emergency programs and services, which may entail [reasonable] modifications to rules, policies, practices, and procedures."

10. "No Charge – People with disabilities may not be charged to cover the costs of measures necessary to ensure equal access and nondiscriminatory treatment."

FIG. 1

Ten federal law key disability nondiscrimination principles.

Credit: U.S. Department of Homeland Security FEMA. Guidance on planning for integration of functional needs support services in general population shelters; 2010 Nov. Available from: https://www.fema.gov/pdf/about/odic/fnss_guidance.pdf. pp. 10–11.

and in professional publications. The main mandate set by all three laws is essentially the same, that disabled people may not "be excluded from participation in or be denied the benefits of the services, programs, or activities of a public entity" [8]. This mandate forms the statutory basis for the disability antidiscrimination requirements this chapter discusses. The main difference between the three laws, for purposes of this chapter, is which entities each law applies to. The ADAA and ADA apply to states and localities (cities, towns, counties), Congress and the federal legislative branch, but not the federal executive branch (federal agencies like FEMA and the US Departments of Homeland Security, Health and Human Services, Justice, and others) [9]. Section 504 applies to federal executive agencies, federal contractors, and entities that receive federal financial assistance. Since many states and localities receive federal funding, this means that they are subject to the ADA, ADAA and Section 504, while FEMA, for example is subject to Section 504, but not the ADA and ADAA. Although the ADA (and ADAA) and Section 504 are separate laws, because the standards set by the laws are essentially the same, they will be treated together in this chapter, unless otherwise specified. The US Department of Justice enforces these laws and individuals can also bring lawsuits under the laws. In ADA Title II litigation against a state or locality, if the state or locality's conduct is deemed wrongful or discriminatory then money damages may be awarded to compensate for a loss, punish a wrongdoer, or discourage future wrongdoing [10].

How the laws define "disability"

The laws define "disability" in a number of ways, and this definition has been modified over time. The ADA's broad definition of disability includes: "a physical or mental impairment that substantially limits one or more major life activities"; "bodily impairment"; a history or record of such an impairment; or being perceived by others as having such an impairment [4,5,11].

> …[M]ajor life activities include, but are not limited to, caring for oneself, performing manual tasks, seeing, hearing, eating, sleeping, walking, standing, lifting, bending, speaking, breathing, learning, reading, concentrating, thinking, communicating, and working.
>
> …[A] major life activity also includes the operation of a major bodily function, including, but not limited to, functions of the immune system, normal cell growth, digestive, bowel, bladder, neurological, brain, respiratory, circulatory, endocrine, and reproductive functions [11,§1202 (2A and B)].

Accessibility requirements for emergency management and public health preparedness

Together the disability antidiscrimination laws prohibit disability discrimination and mandate the accessibility of public programs, services and activities, which would include emergency management and public health on the federal, state, and local levels [12,13]. These laws require that emergency management and public health programs, services and activities provide: (1) physical accessibility; (2) program access (equal access); (3) provision of "reasonable modifications" as needed; and (4) "effective communication," including "auxiliary aids and services" as needed; (5) in the most integrated setting. Fig. 2 illustrates these five requirements [7].

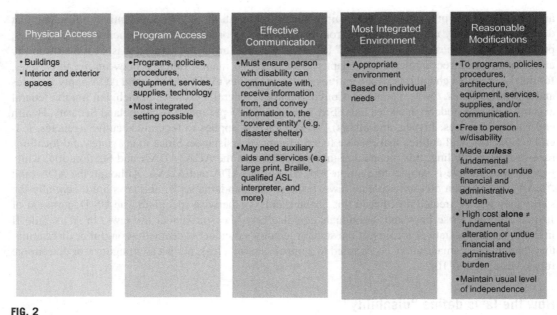

FIG. 2

Summary of antidiscrimination standards.

Physical accessibility and program accessibility

"Physical access" refers to buildings and includes interior and exterior spaces. An example of a physical access issue would be whether a building has an entry ramp or an accessible restroom [13]. Accessibility is usually required for the following areas: "parking, drop-off areas, entrances and exits, security screening areas, toilet rooms, bathing facilities, sleeping areas, dining facilities, areas where medical care or human services are provided, and paths of travel to and from and between these areas" [7, p. 10].

"Program access" refers to equal access to programs, policies, procedures, equipment, services, supplies, and technology [13]. These include, for example, "emergency preparedness, notification of emergencies, evacuation, transportation, communication, shelter, distribution of supplies, food, first aid, medical care, housing, and application for and distribution of benefits" [7]. So, if a local government-run emergency shelter holds a children's story time, for example, then there must be a way for a Deaf child to be able to access the program. This might mean that an American Sign Language (ASL) interpreter, paid for by the local government, signs as someone else reads the book out loud. Or this might mean that a child with autism who doesn't like to feel crowded, sits apart from the other children while the story is being read. If the shelter posts notices or makes announcements, program access (and effective communication) might mean that notices are posted on the shelter's bulletin board, shared over a public address system, and repeated daily by someone who is certified as a sign language interpreter.

Reasonable modifications

"Reasonable modifications" are adjustments or accommodations in programs, services, activities, or procedures that allow a person with a disability to participate [14]. Reasonable modifications may include Braille or large print emergency shelter registration forms, or a quiet space at a vaccination site to decrease sensory stimulation for someone with a sensitivity to noise or crowds. Or suppose the local disaster shelter is crowded and there is little space in the aisles. A likely reasonable modification for someone who uses a wheelchair would be for shelter personnel to keep the aisles clear by re-arranging the space. Alternatively, as a reasonable modification shelter personnel might provide wayfinding support (guiding someone to navigate an environment [15]) to orient a person with an intellectual disability to the shelter layout and set up. Per the US Department of Justice, reasonable modifications in disaster shelters include, but are not limited to, modified shelter cots, extra storage for durable medical or other equipment, access to the shelter kitchen if needed for medical reasons (such as for someone who requires blended food because of swallowing problems), help to complete forms, and personal assistance services, all without charge [7].

Personal assistance services are provided by people (either paid or unpaid) to help disabled people with daily living activities like bathing, eating, dressing and the like. Personal assistance services in an emergency shelter must be available for those who usually receive them. The idea is that the program and services must enable the individual with a disability to maintain their usual level of independence [16]. The potential barriers to providing these services in an emergency shelter include cost and finding qualified service providers. There may be some practical ways to potentially limit the extent of the challenge. Fig. 3 illustrates one such strategy and following Fig. 3 there is a discussion of other strategies. The goal of advance planning is to minimize the potential challenge of a reasonable modification as much as possible, with the understanding that it may not be feasible to make the challenge go away.

Personal Assistance Services

"Local governments and emergency shelter operators may *not* require someone to bring their own direct support professional in order to receive [disaster] shelter services. But local plans [likely] *could* call for publicizing that direct support professionals are welcome in the local emergency shelter.

"In engaging with [emergency and public health preparedness] planners, it's important to emphasize not just the law, not just that something is the right thing to do, but that there are practical benefits too. For example, publicizing that direct support professionals are welcome might cut down on the number of people who arrive at an emergency shelter without a direct support professional. Fewer people who need personal care assistance might mean lower costs and less work for shelter staff."

Credit: Association of University Centers on Disabilities (AUCD). Prepared4ALL Whole Community Inclusive Emergency Planning [Internet]. Association of University Centers on Disabilities (AUCD). 2021 [cited 2021 Oct 21]. Available from: https://nationalcenterdph.org/our-focus-areas/emergency-preparedness/prepared4all/online-training/. Used with permission. [16]

FIG. 3

Personal assistance services: one strategy to address the issue.

Credit: Association of University Centers on Disabilities (AUCD). Prepared4ALL whole community inclusive emergency planning. Association of University Centers on Disabilities (AUCD); 2021. [cited 2021 Oct 21]. Available from: https://nationalcenterdph.org/our-focus-areas/emergency-preparedness/prepared4all/online-training/. Used with permission.

Note that in order to implement this strategy there would need to be advance planning. The strategy suggested in Fig. 3 arose during an inclusive emergency planning meeting held in Massachusetts (MA). There are, however, other strategies that might be employed to address this need. For example, if an individual's personal care provider doesn't want to remain in the shelter while providing services, another simple and cost free strategy would be for shelter policy to allow the provider to come and go as needed to provide their services to their client. This would involve modifying a shelter's "no visitor" policy. On the local level, as the need for these services exceeds capacity, the locality can apply to the state for help. And, if the need exceeds the state's capacity during a presidentially declared disaster, the Governor may request these services from FEMA [17]. As part of a local advance planning process, emergency management and public health personnel might research requirements for requesting FEMA funding for personal care assistance so they would be ready to proceed with such a request if needed during an emergency.

It bears repeating that there may be more than one way to provide reasonable modifications and remain legally compliant. There is no one list of all possible reasonable modifications because these modifications are based on individual need and the circumstances. And reasonable modifications need not necessarily be expensive. For example, in a 2020 survey about workplace reasonable accommodations (analogous in many ways to reasonable modifications) results indicated that 56% of workplace accommodations cost nothing and the average cost of the remaining 44% was about $500 [18]. Additionally, it is important to remember that reasonable modifications need not always involve a purchase; for example, equipment may be borrowed, rented or bartered. Simple adjustments to current equipment, emergency services, or environments may be easy to make. For example, in one MA inclusive emergency planning meeting a collaborative group of community members and local government officials determined a need for large print versions of key disaster shelter forms. A question was raised about what the plan would be if the large print copies didn't suffice. The group knew that the public library had CCTV devices (video magnifiers) for enlarging the print in books. A no cost solution for large print needs would be to borrow a CCTV from the library for use in the disaster shelter (assuming there was power). The group's action plan called for a meeting with the library to discuss this possibility.

Effective communication

Effective communication is critical to support emergency management throughout the disaster life cycle. Specifically, "effective communication" means that "[a] public entity must ensure that its communications with individuals with disabilities are as effective as communications with others" [19]. To provide effective communication, the state or locality must make appropriate "auxiliary aids and services" available when needed. "Auxiliary aids and services" include ways to make visual materials accessible to Blind people by using Braille translations, audio recordings, or other means. They also include the availability of a sign language interpreter for a Deaf person who needs one during a meeting or providing "plain language" (simplified text) emergency shelter registration forms. But auxiliary aids and services aren't limited to Deaf or Blind people. An auxiliary aid or service might include having a shelter registration form read aloud to someone with a cognitive disability to increase comprehension. Both reasonable modifications and auxiliary aids and services are required unless the potential modification or aid would change the fundamental nature of the service, program, or activity or pose undue (significant) financial and administrative burdens [19]. What is a "fundamental alteration," or "undue financial and administrative burden" is dependent on the facts of the situation. High cost alone does not make a modification unreasonable [20]. The law does not include a specific list of reasonable modifications, but rather general principles. Sometimes a determination about reasonableness must be decided by a court due to litigation after the fact. There is discussion about this topic below. A person with a disability may not be charged for the cost of a reasonable modification, auxiliary aid, or service. A list of some auxiliary aids and services that might be needed can be found in Table 1.

In addition to the auxiliary aids and services already mentioned, there may be other ways to address communication needs and ensure effective communication. For example, darkness makes faces and hands difficult to see and impedes communication for those who use visual communication methods, such as a Deaf person who lip reads and signs. One example of a potential strategy for an emergency responder to use when communicating at night with someone who signs (when a sign language interpreter is not available), is to shine a flashlight on their own face while they speak. This would illuminate the responder's face so that the Deaf person could lip read. For two way communication, Deaf people might text back and forth with the first responder to facilitate understanding. A simple dry erase board may also enable short, quick communications in specific situations. The responder would need to keep sentences short and simple, because someone fluent in ASL may not be fluent in English as they two are different languages. See Fig. 4 for more about sign language. While the law requires that auxiliary aids and services are provided, multiple strategies to facilitate effective communication exist and what is appropriate depends on the needs of the individual with a disability and the specific circumstances at hand.

Table 1 Auxiliary aids and services-examples.

1. Written materials
2. Note takers
3. Exchange of written notes
4. Qualified sign language interpreters
5. Assistive listening devices
6. Audio recordings
7. Hearing aid compatible phones
8. Text messaging
9. Video-based communication and interpreting systems
10. Plain language (simplified) materials
11. Large print materials
12. Qualified readers
13. Braille materials
14. Email

Note the emphasis on "qualified" sign language interpreters and readers and that the aids and services are not listed in order of importance or priority.
Credit: Association of University Centers on Disabilities (AUCD). Prepared4ALL whole community inclusive emergency planning. Association of University Centers on Disabilities (AUCD); 2021 [cited 2021 Oct 21]. Available from: https://nationalcenterdph. org/our-focus-areas/emergency-preparedness/prepared4all/online-training/; ADA National Network. Effective communication. ADA National Network; 2017 [cited 2022 Jan 2]. Available from: https://adata.org/factsheet/communication. Used with permission.

American Sign Language

- American sign language (ASL) is a different language than written or spoken English. Someone who uses ASL to communicate may not have the same level of English fluency.
- A sign language interpreter is an auxiliary aid or service under the ADA.
- A sign language interpreter is not required for all emergency communication situations. Sign language interpreters may be required when the information being communicated is complex or goes on for a long period of time. Factors to consider can include the context of the conversation, the number of people involved, and the importance of the communication.
- Emergency plans may not require someone who communicates via ASL to bring an interpreter with them.
- An accompanying adult or child cannot be required to act as an interpreter, except in emergencies 'involving an imminent threat to the safety or welfare of an individual or the public where there is no interpreter available.'
- A person's usual communication method is best, but there may be circumstances in which that method is not feasible or available in reality (ie. not merely a matter of convenience). Then factors like the context and importance of the communication, communication environment, extended or limited timeframe for the communication, the topic and its importance and the length of the communication become important to determine the communication method to use.

FIG. 4

Description of American Sign Language.

Credit: Association of University Centers on Disabilities (AUCD). Prepared4ALL whole community inclusive emergency planning. Association of University Centers on Disabilities (AUCD); 2021. [cited 2021 Oct 21]. Available from: https://nationalcenterdph.org/our-focus-areas/emergency-preparedness/prepared4all/online-training/. Used with permission.

Services delivered in the most integrated manner

Emergency services, activities, and programs must be delivered in the "most integrated setting appropriate to the needs" of individuals with disabilities [21]. This is an integration mandate and, for example, means that generally individuals with disabilities must be served in the same disaster shelters as everyone else, with reasonable modifications as needed, even if they need personal assistance services [7,21]. In addition to separating people with disabilities from friends, family, caregivers and their usual supports, FEMA notes that inappropriate placement in a "medical" or "special needs" shelter may create inefficiencies and jeopardize the well-being of all community members by causing unnecessary use of emergency medical resources. The integrated setting requirement also means that most disabled people should have access to healthcare services such as flu and COVID-19 vaccines in the same places as people without disabilities and should receive other emergency, disaster, healthcare, and public health services in the most integrated environment as well.

Service animals and comfort animals

The ADA has specific requirements related to service animals. The law requires that service animals accompany their owners almost everywhere the owner can go including evacuation vehicles, emergency vaccine or medication dispensing sites, emergency shelters, and similar areas [22]. Service animals are working animals trained to perform tasks to assist people with disabilities. Examples include assistance dogs to lead Blind people around obstacles or dogs that can detect seizures in someone with epilepsy. The ADA definition of service animal only includes dogs or miniature horses, but not other animals [22]. While the ADA does not apply to comfort animals, whose role is to provide comfort or emotional support without being specially trained to do so, local or state law may permit comfort animals the same access as service animals. Service animals have no requirement for registration or special tags.

Emergency and disaster personnel may not ask the animal's owner if they have a disability or about the nature of their disability. However, they may ask the following two questions of the owner regarding the use of the service animal: (1) Is this a service animal required because of a disability? (2) What tasks has the animal been trained to perform? [22]. While it is the owner's job to care for the service animal, emergency and disaster personnel may ask the owner questions about the animal's needs without asking specific questions about the disability itself.

The service animal must remain under the owner's control and the owner is required to provide care or food for their service animal. Emergency responders, shelter staff, or emergency management and public health preparedness personnel are not expected to be responsible for maintaining the service animal [22]. A person with a disability cannot be asked to remove their service animal unless the animal is not house broken, or the owner can't control the animal. As with other reasonable modifications, people with disabilities who use service animals cannot be isolated from other people in a disaster shelter or charged fees specifically for having a service animal [23].

Key actors at the local and state levels

After focusing on accessibility requirements, it is important to examine key actors at the state and local levels and their roles related to the fulfillment of these requirements. At the local level, key actors responsible for accessibility and inclusion implementation include county, city, and town public health and emergency management personnel, first responders, and emergency shelter staff (whether or not they are paid or volunteers [13]) and others responsible for the delivery of emergency, disaster, and pandemic services, shelter admission and reasonable modification decisions. At the state level, the key actors are state government personnel, such as the directors of the state emergency management and public health agencies and those who work for them. If the local government contracts with a third party for services or programs, then the contractor must also follow the ADA requirements. For example, if a county hires the American Red Cross or another non-profit organization to develop and manage shelter operations, then that non-profit organization is bound by the ADA's requirements and needs to provide physical and program access, and effective communication, with reasonable modifications and auxiliary aids for people with disabilities as needed [13].

History of persistent challenges with ADA compliance

Although the ADA was passed in 1990, and Section 504 before that, implementation related to emergency management has been inconsistent [24]. There are numerous examples which illustrate the need for better understanding of the requirements and highlight areas for improvement. During Superstorm Sandy (2012) and earlier events, people with disabilities reported instances of inaccessible shelters [25], denial of shelter admission [26] and lack of reasonable modifications such as the availability of sign language interpretation [25]. In other situations, people with disabilities who had no significant medical needs reportedly were sent to nursing homes and other institutional settings absent the need for that level of care [27,28]. Others reported that they were not allowed to leave their disaster shelter or access their own funds while in shelters, likely based on misconceptions of their disability by shelter personnel [27]. The same or similar issues were reported in 2017–2018 by the Partnership for Inclusive Disaster Strategies, a disability-led organization. For example, their report describes a number of exclusionary practices including unnecessary institutionalization due to lack of appropriate housing availability or challenges faced navigating the disaster recovery process [29].

Project Civic Access

Project Civic Access (PCA) compliance reviews are formal reviews of local jurisdiction compliance with the ADA instituted by the US Department of Justice (DOJ) [30]. The review process examines local emergency management compliance with the ADA, among other areas of compliance. The reviews may be instituted either based on local ADA-related complaints or initiated by the DOJ. According to the DOJ, most jurisdictions are aware of their ADA obligations and fully cooperate with these reviews [30]. Outcomes of the reviews have resulted in settlements that promote and guide inclusive planning and provide key lessons for other communities.

A PCA review for the city and county of Denver, Colorado in 2018 (updated in 2021) [31], resulted in the city and county agreement to: (1) develop a process to seek and incorporate public input into the local emergency plan; (2) develop an equitable evacuation plan and provide access to appropriate accessible post-emergency housing; (3) provide accessible warning communication systems; (4) assure accessible emergency shelters with back-up power and refrigeration for medications; and (5) develop a policy to allow disabled people to keep their service animals with them [31]. In 2021 the DOJ deemed the county to be ADA compliant, having rectified the areas of non-compliance identified in the 2018 review [32]. Lessons learned, which could be replicated elsewhere, include the importance of seeking public input into emergency plan development, equity in the evacuation process, communicating disaster warnings, accessible post-disaster housing, and that reasonable modifications in emergency shelters should include backup power and medication refrigeration.

In 2020 the DOJ Project Civic Access settlement agreement with the city and county of Killeen, Texas called for the same remediation as above, plus additional actions [33]. The settlement called for the emergency plan to specifically state that Killeen requires emergency services to be provided in the most integrated setting appropriate for people with disabilities and must describe specific methods to achieve integration. Killeen is also to plan for accessible post-disaster temporary housing in the most integrated setting to avoid even temporary housing in institutional settings [33]. The reality for many people with disabilities is that a disaster destroys their already accessible home. In most communities generic accessible housing is limited. Additionally, many people with disabilities who require accessible housing have had home modifications completed through certain Medicaid programs which enable modifications specifically tailored to meet their individual needs [34]. Also see [35], where home modifications are included under "other services." Both of the settlements areas for emergency management and public health emergency preparedness to pay attention to and address.

Federal litigation challenging the adequacy of local emergency plans

Changes in emergency management practices may come about in different ways, including policy decisions initiated by a local government agency or the legislative arm of the government, due to public pressure, or via various legal proceedings. As previously seen, PCA brings about systems change after an investigation has found ADA compliance gaps and a settlement agreement for remediation has been agreed to. Systems change can also result from legal actions brought in court, whether they are resolved by a judge, jury, or via a settlement between the parties. Class action lawsuits are a particular type of legal action where one or a few parties represent the interests of a larger group of people with the same legal dispute. This section summarizes the issues raised in California, New York City and Washington,

DC class action lawsuits alleging that local emergency plans inadequately provided for people with disabilities and were therefore discriminatory. In the cases summarized below the court either found disability discrimination or the parties negotiated a settlement before the court rendered a decision likely to be a finding of discrimination.

The ADA and other disability antidiscrimination laws are explicit in their unwavering protection of individuals with disabilities. However, there is no specific instruction in the statutes about how these rights mesh with and should be incorporated into emergency management and public health preparedness planning [36]. Similar to the Project Civic Access settlements, statements from the court orders, trials, and settlement agreements can be a useful guide to incorporating disability rights into emergency management and public health preparedness. The inadequacies of the emergency plans at issue in the New York City, Washington, DC and California cases indicated the same gaps noted in the scholarly and gray literature (see Chapter 4) and Project Civic Access, such as shelter, evacuation, transportation and communication inaccessibility. Additionally, in the New York City case the judge found no advance planning for power outages, high rise building evacuations and large scale events [25].

The judge's Opinion and Order in the New York City case provides a strong statement about the aims of disability antidiscrimination laws, and explains that these laws provide protection not just against intentional disability discrimination, but also "discrimination that results from 'benign neglect'…" [legal citations omitted] [25] and that the government's responsibility is to "do more than provide a program on equal terms to those with and without disabilities; they require 'affirmative accommodations to ensure that facially neutral rules do not in practice discriminate against individuals with disabilities.'" [legal citation omitted] [25].

The court in the Los Angeles case stressed the importance of thoughtful advance planning rather than ad hoc decision making [36]. A plan to respond to the needs of people with disabilities on an as needed basis is

> both legally inadequate and practically unrealistic…the purpose of the City's emergency preparedness program is to anticipate the needs of its residents in the event of an emergency and to minimize … last-minute, individualized requests for assistance described by the City, particularly when the City's infrastructure may be substantially compromised or strained by an imminent or ongoing emergency or disaster [37, p. 25, 26]

The settlement agreements in the New York City and Los Angeles cases illustrate specific steps for emergency management and public health departments to take to protect people with disabilities, along with the rest of the community, from the impacts of emergencies and disasters. For example, the agreements call for hiring a disability/disaster expert to consult to improve the local emergency plan [25,37]. The settlements require creating new positions within the local emergency management structure, such as a Disability and Access and Functional Needs Coordinator, with sufficient authority to successfully implement change, and an Advisory Panel to provide strong input into planning beyond giving advice [25]. Other lessons learned from the settlements include the importance of advance communication planning which should include a strategy to address service outages, systems failures, and accessible warning systems, and a plan to maintain a shelter accessibility database (see, e.g., Ref. [38]). Furthermore, the agreements call for accessible transportation and evacuation policies that do not solely rely on public transport and include strategies for high rise building evacuation planning [25]. The Los

Angeles case and the Washington, DC case focus additional attention on disaster shelters. Shelter services and activities must be accessible and inclusive, with plans to address medication and supply chain disruption and to replace lost or damaged medical equipment [39,40]. And the post-disaster recovery phase should include a process for post-disaster rapid community needs assessment [25].

Litigation related to the effective communication mandate

Litigation outcomes related to the effective communication requirement can also guide emergency management and public health preparedness. This mandate is critical in emergencies and disasters, because without it a person with a disability may not be able to understand and follow emergency instructions, evacuate, or develop an emergency plan, to name a few. Two legal actions shed light on what makes communication "effective." In the first case several people who were Deaf or Hard of Hearing and a national Deaf advocacy organization sued the state of Arizona, three cities, and a county alleging disability discrimination because the local emergency 9-1-1 service wasn't accessible 24-7 to them [41, 42]. The technology Arizona chose for 9-1-1 was only accessible to Deaf people and those who are Hard of Hearing if used in conjunction with a TTY machine (a device that adapts a telephone) or Telecommunication Relay System (TRS) (where an operator types what someone says and reads what someone types) with high speed internet or a hotspot. These technologies were generally only available to the people affected when they were at home. This situation created an obvious inequity and was not effective communication. People who were Deaf or Hard of Hearing had access to the 9-1-1 system only at home, whereas other Arizonans had access no matter their location. The legal issue was resolved when the state launched its *Text-to-911* program which enabled 9-1-1 services to receive texts from cell phones, which many people who are Deaf or Hard of Hearing carry [41,42]. One lesson to be learned from this case is about the benefits of universal design. The *Text-to-911* program uses technology that is accessible to everyone with a cell phone, whether or not they have a disability. If this technology had been chosen initially, with input from people with hearing conditions, there would have been no legal issue, a benefit of inclusive advance planning.

In the second legal action, related to a hazardous chemical spill in Minnesota (MN), the question was whether the optimum communication modality is always the one that must be used, particularly in situations where decisions must be made very quickly. In other words, does effective communication require only the most favorable communication modality? The case involved a chemical spill that exposed 49 people plus vehicles and homes to mercury [43]. Four of the 49 people who needed to be decontaminated were Deaf people who communicated via ASL. No sign language interpreter was available during the decontamination process and communication occurred through a number of modalities including writing notes, hand gestures, lip reading, and limited sign language. The issue before the court was whether there had been "effective communication" during the decontamination process because an ASL interpreter was not provided. The court found that, under the specific circumstances of the situation, there was effective communication and therefore no unlawful discrimination. The court reasoned that the evidence showed that the Deaf people had all successfully completed the decontamination process by using the multiple communication modalities mentioned above plus oral communication with hearing family members at the scene [43]. The chronology of events showed that people had to be decontaminated quickly and there wasn't time to bring a qualified sign language interpreter to the scene, making alternatives necessary. To support its reasoning, the court acknowledged that the communication means used were not perfect, but stated that it did not put the four Deaf people in a different position than the hearing individuals. In other words, the court found that under the facts of this situation, communication did not have to be optimum to be effective [43]. Note that under other circumstances, a court might not consider these same communication methods to be effective.

Equity during medical resource shortages

Up to this point in the chapter the discussion has focused on natural and human-caused disasters, but disability antidiscrimination laws apply equally to pandemics. Under the ADA and other laws people with disabilities are entitled to equal access to COVID-19 services and treatments, effective communication from public entities, and physically accessible places in which to receive vaccines and medical treatment.

In situations where there isn't enough medication, vaccines or healthcare available for everyone who needs them, hospitals and healthcare providers may be forced to decide who should receive care and who shouldn't [44]. At such times, the usual medical standards of care may be altered. "Crisis standards of care" are the standards that decision makers will use to determine who gets what care when medical resources are limited. Since there are no uniform federal guidelines for crisis standards of care, each state adopts its own policies [45]. However, making these decisions based on disability violates Section 504, the ADA and the Affordable Care Act (Section 507). During the COVID-19 pandemic a number of complaints related to this issue were filed with the US Department of Health and Human Services Office of Civil Rights. For example, Alabama's crisis standards of care included a standard for rationing ventilators that discouraged providing ventilators to "persons with severe mental retardation" [46]. In 2020 the Office for Civil Rights resolved this issue with the state of Alabama and state's crisis standards changed. The law is clear that:

Health care providers can't limit or deny care to a person with a disability because of their disability or need for reasonable modifications. They can't limit or deny care because a person with a disability may require more intensive care or is less likely to survive. Providers should provide the [same level of] medical care to people with disabilities that they provide to anyone else [16]

Other civil rights statutes

This section changes focus to look at antidiscrimination laws other than those relating to disability antidiscrimination. This is important because disabled people may also identify as members of other historically marginalized groups. This concept is known as "intersectionality," because a person's identities in more than one marginalized group may intersect. In that case, in addition to the ADA and Section 504, there are other important antidiscrimination laws that would apply in emergencies and disasters.

The Stafford Act is the legal source for the President's power to make federal disaster declarations and is the law that permits federal disaster assistance to go to the states [47]. The Stafford Act provides antidiscrimination protection for a broad group of people. "[The] distribution of supplies, the processing of applications, and other relief and assistance activities shall be accomplished in an equitable and impartial manner, without discrimination on the grounds of race, color, religion, nationality, sex, age, disability, English proficiency, or economic status" [47,§308]. The Act extends these antidiscrimination requirements to private disaster relief organizations, which means that a private disaster relief organization is held to the same antidiscrimination standards as the government.

Another statute with discrimination protections is the Patient Protection and Affordable Care Act of 2010 (ACA). Section 1557 of the ACA prohibits discrimination on the basis of race, color, national origin, sex, age, or disability in health programs or activities run by the federal government or which receive federal financial assistance [48]. This law would apply to public health preparedness activities and to disaster-related health services, programs, and activities which received federal funding (like emergency shelter services).

The Civil Rights Act of 1964, title VI, also applies in emergencies. It prohibits discrimination on the basis of race, color, or national origin (including limited English proficiency) [49]. To address race discrimination during emergencies, in 2016 the US Department of Justice and other federal agencies issued joint guidance to recipients of emergency management federal financial assistance, reiterating that Title VI prohibits federal financial recipients from unlawful discrimination during emergencies and disasters [50]. The law includes prohibitions against intentional discrimination and facially neutral policies and practices with discriminatory impact. These practices and policies do not have explicitly discriminatory actions or language, but in fact have a discriminatory impact, application or affect. The joint guidance document states that public engagement with diverse populations is a "critical step" in assuring non-discrimination [50].

Waiving HIPAA and other laws during a disaster or emergency

Another legal issue that arises in emergencies and disasters is whether or not laws can be waived (suspended and considered not to be in effect for a period of time) due to such events. In general, US laws are not waived in the ordinary course, but some laws can be suspended in limited emergency situations. Orenstein notes that, for example, laws related to a governor's authority can be waived during emergencies if permitted by state statute [51]. At least one research team found that there seems to be some confusion about this concept among emergency managers, but not public health personnel. Botoseneanu and colleagues found that the emergency managers they studied felt that many laws could be waived due to the exigencies of an emergency. The research team highlighted the following quotes from emergency managers: "… everything is based on laws, but laws can be put on hold to get it done, get people to safety whatever it takes …" and " … [we] can go outside the legal apparatus to get what's needed …" [52, p. 364]. The researchers did not have the same finding related to public health professionals; the public health professionals they studied were more concerned about the exact nature of their legal authority and how to comply with that authority.

Limited waivers of some provisions of the federal Health Insurance Portability and Accountability Act of 1996 (HIPAA) privacy and health information security law and Medicare and Medicaid regulations are allowed under some emergency circumstances. HIPAA's aim is to protect people's privacy related to certain personal health information, called Protected Health Information or PHI, and the law governs how and with whom this information may be shared by healthcare providers such as clinicians, hospitals, nursing homes, clinics, health plans, etc. [53]. The standard the law follows is one of balancing the patient's interest in privacy versus public health needs [54]. In other words, there are certain situations in which it is permissible to share patient information without the patient's permission for the greater good. Examples include: to treat the patient, for public health purposes (for example if a public health authority needs the information as part of a data collection effort to prevent disease or where there is an impending public health threat), to family and friends involved in patient care, and to persons at risk of contracting or spreading disease if otherwise permitted by law [54].

While the HIPAA rule cannot be suspended in its entirety, under the law the US Secretary of Health and Human Services may waive specific provisions of the law [54,55]. In order for this to happen the President must declare a state of emergency or disaster, the US Secretary of Health and Human Services must declare a public health emergency, and the hospital or health facility at issue must have enacted its disaster protocol. The waiver of these provisions is limited to the area covered by the emergency and the waiver has a duration of 72 h [56].

In March 2020, during the COVID-19 pandemic, the US Secretary of Health and Human Services used the Project Bioshield and Social Security Acts to waive some of HIPAA's sanctions and penalties due to the declared COVID-19 health emergency [55]. The Secretary waived HIPAA sanctions for communicating with a patient's family or friends involved in their care without prior patient consent and failing to provide a patient with the HIPAA privacy notice, honor patient requests for certain privacy protections, confidential communications, and the right to opt out of a healthcare facility patient name directory [55].

The Centers for Medicare & Medicaid Services also waived portions of the Medicare and Medicaid regulations due to the pandemic, for example allowing patient transfers directly to a nursing home after a 3 day hospital stay during a health emergency [57]. While the intent was to help Medicare and Medicaid recipients access care more easily, an unexpected consequence was to make it easier for someone with a disability who was hospitalized to be transferred to a nursing facility rather than back to their home in the community. This issue has been reported in previous disasters as well [28].

On the other hand, civil rights laws, such as the ADA, Section 504, Section 1557 of the Affordable Care Act, and the Civil Rights Act of 1964, are never waived. In its statement on Civil Rights and COVID-19, the US Department of Justice reaffirmed its commitment to robust enforcement of civil rights laws during the COVID-19 pandemic and re-stated that antidiscrimination protections cannot be waived even during emergency events.

> The COVID-19 pandemic has stressed our Nation's commitment to an open, equal, and inclusive society. We have seen hateful and xenophobic rhetoric and violence aimed at Asian American and Pacific Islander (AAPI) communities and businesses. We have also seen Black, Indigenous, Latino, and Pacific Islander communities, as well as people with disabilities, suffer disproportionately high rates of death and greater risk of infection and hospitalization....The Department of Justice will vigorously enforce Federal civil rights as we continue the process of national reckoning, recovery, and healing. Civil rights protections and responsibilities still apply, even during emergencies. They cannot be waived [58].

Conclusion

FEMA provides a lengthy inclusive emergency management philosophy statement in its whole community guidance [59], the US Department of Health and Human Services Office of the Assistant Secretary for Preparedness and Response (ASPR) and the CDC refer to this concept in their capabilities

[60,61] and, as discussed in this chapter, there are a number of relevant federal civil rights and disability antidiscrimination statutes which promote equity and inclusion. While there are some guidance documents with practical advice, such as the US Department of Homeland Security FEMA's Guidance on Planning for Integration of Functional Needs Support Services in General Population Shelters [7], there is little government guidance that provides specifics about inclusive emergency management implementation. To expand on practical guidance, one may look to the legal settlements from Project Civic Access, class action and other lawsuits for useful implementation examples. It is clear that civil rights protections can never be waived during emergencies, disasters or pandemics, although apparently there is confusion about that issue among some professionals in the field [52]. Jacobson and colleagues' [1] research indicates that both local public health practitioners and emergency managers don't accurately understand the law and base decisions on perceptions rather than actual legal requirements. Sherry and Harkins [24] concur that emergency managers are confused about legal requirements; they argue that non-compliance may be caused by concerns about unrealistic legal requirements. The question remains, as to the best way to strengthen emergency plans and practices in order to expand disability access and equity in emergencies, disasters and pandemics. The answer may lie in learning from court decisions and legal settlements and from promising practices as described in Chapter 10.

Questions for thought

(1) The Americans with Disabilities Act is a very well-known law which was passed more than 30 years ago. As with other well-known antidiscrimination laws, disability discrimination remains despite the law's existence. Why do you think disability discrimination in disasters and emergencies continues, despite the statutes, the lawsuits and Project Civic Access?

(2) Consider the following policy:

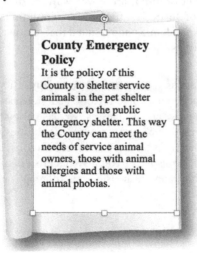

County Emergency Policy

It is the policy of this County to shelter service animals in the pet shelter next door to the public emergency shelter. This way the County can meet the needs of service animal owners, those with animal allergies and those with animal phobias.

(a) What are the likely practical consequences of this policy?

(b) Is this policy an example of whole community planning? Is the policy in accordance with the Americans with Disabilities Act (ADA)?

(c) What are the strongest ADA-related arguments in favor of and against this policy?
(3) Consider Fig. 5 below, adapted from Guidance on Planning for Integration of Functional Needs Support Services in General Population Shelters [7]. The topics in the graphic are essentially the same ones mentioned at the beginning of the chapter in Fig. 1. Think of one example to illustrate each concept in the outer circle.

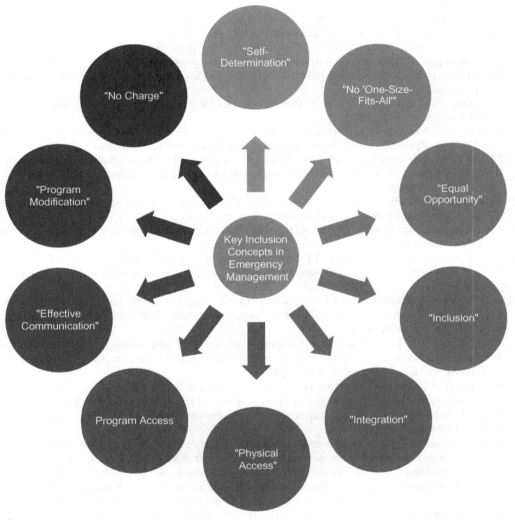

FIG. 5

Key inclusion concepts in emergency management.

Credit: Adapted from U.S. Department of Homeland Security FEMA. Guidance on planning for integration of functional needs support services in general population shelters; 2010 Nov. Available from: https://www.fema.gov/pdf/about/odic/fnss_guidance.pdf.

pp. 10–11.

References

[1] Jacobson PD, Wasserman J, Botoseneanu A, Silverstein A, Wu HW. The role of law in public health preparedness: opportunities and challenges. J Health Polit Policy Law 2012;37(2):297–328.

[2] U.S. Constitution 10th Amendment.

[3] Ransom MM, Valladares LM, editors. Public health law: Concepts and case studies. Springer Publishing Company; 2021 Jul 15.

[4] Americans with Disabilities Act of 1990, as amended, 42 U.S.C. §§ 12101 et seq.

[5] Americans with Disabilities Act Amendments Act of 2008, 42 USC 12101.

[6] Section 504 of the Rehabilitation Act of 1973, as amended, 29 U.S.C. § 794.

[7] U.S. Department of Homeland Security FEMA. Guidance on planning for integration of functional needs support services in general population shelters, 2010 Nov. Available from: https://www.fema.gov/pdf/about/odic/fnss_guidance.pdf.

[8] Americans with Disabilities Act Title II Regulations. 28 CFR § 35.130. Available from https://www.law.cornell.edu/cfr/text/28/35.130.

[9] ADA National Network. Is the Federal government covered by the ADA? ADA National Network; 2022. [cited 2022 Jan 8]. Available from: https://adata.org/faq/federal-government-covered-ada.

[10] DBTAC Southwest ADA Center at ILRU. Remedies under the ADA. DBTAC Southwest ADA Center at ILRU; 2011. [cited 2022 Dec 21]. Available from: http://www.southwestada.org/html/publications/ebulletins/legal/2010/june2010.pdf.

[11] Americans with Disabilities Act of 1990, as amended, 42 U.S. C. § 12102 (2A and B)—Definition of disability. Available from https://www.law.cornell.edu/uscode/text/42/12102.

[12] Americans with Disabilities Act of 1990, 42 U.S.C. § 12131(2). Available from: https://www.law.cornell.edu/uscode/text/42/12131.

[13] U.S. Department of Justice. Chapter 7 Emergency management under Title II of the ADA. In: ADA best practices tool kit for state and local governments; 2007. Available from: http://www.ada.gov/pcatoolkit/toolkitmain.htm.

[14] U.S. Department of Justice Civil Rights Division Disability Rights Section. Americans with Disabilities Act ADA update: A primer for state and local governments. Available from: https://www.ada.gov/regs2010/titleII_2010/title_ii_primer.html.

[15] Montgomery County Department of Transportation. Improving wayfinding and safety for people with a vision disability. Available from: https://www.montgomerycountymd.gov/DOT/Resources/Files/VisionDisability-UrbanNavigationStudy_MeetingPresentation.pdf.

[16] Association of University Centers on Disabilities (AUCD). Prepared4ALL whole community inclusive emergency planning. Association of University Centers on Disabilities (AUCD); 2021. [cited 2021 Oct 21]. Available from: https://nationalcenterdph.org/our-focus-areas/emergency-preparedness/prepared4all/online-training/.

[17] FEMA. Addendum to the mass care/emergency assistance pandemic planning considerations: Delivering personal assistance services in congregate and non-congregate sheltering, 2020 Dec. Available from: https://www.fema.gov/sites/default/files/documents/fema_personal-assistance-services_addendum_12-07-2020.pdf.

[18] Roussey B. Average costs of reasonable accommodations in the workplace, 2021. [cited 2022 Jan 2]. Available from: https://www.accessibility.com/blog/average-costs-of-reasonable-accommodations-in-the-workplace.

[19] U.S. Department of Justice. The Americans with Disabilities Act Title II technical assistance manual covering state and local programs and services. Available from: https://www.ada.gov/taman2.html.

[20] ADA National Network. Effective communication. ADA National Network; 2017. [cited 2022 Jan 2]. Available from: https://adata.org/factsheet/communication.

[21] U.S. Department of Justice, Civil Rights Division, Disability Rights Section. Statement of the Department of Justice on enforcement of the integration mandate of Title II of the Americans with Disabilities Act and Olmstead v. LC, 2020 Feb. Available from: https://www.ada.gov/olmstead/q&a_olmstead.htm.

[22] Rocky Mountain ADA Center. Service animals. Rocky Mountain ADA Center; 2020. [cited 2022 Jan 2]. Available from: https://rockymountainada.org/topics/individuals/service-animals.

[23] U.S. Department of Justice Civil Rights Division Disability Rights Section. ADA requirements service animals, 2020 Feb. Available from: https://www.ada.gov/service_animals_2010.htm.

[24] Sherry N, Harkins AM. Leveling the emergency preparedness playing field. J Emerg Manag 2011;9(6):11–5.

[25] Brooklyn Center for Independence of the Disabled (BCID), et al. v. Bloomberg, No. 11 Civ. 6690 (JMF) (S.D.N.Y. Nov. 7, 2013)., 2013. Available from: https://dralegal.org/case/brooklyn-center-for-independence-of-the-disabled-bcid-et-al-v-mayor-bloomberg-et-al/.

[26] Jan S, Lurie N. Disaster resilience and people with functional needs. N Engl J Med 2012;367(24):2272–3. https://doi.org/10.1056/NEJMp1213492.

[27] Testimony of Shelley Weizman. Oversight: Emergency planning and management during and after the storm: Emergency preparedness and response at the city's healthcare facilities: Hearing before the New York City Council Committee on Mental Health, Mental Retardation, Alcoholism, and Drug Abuse and Disability Services Committee on Aging., 2013 Jan. Available from: http://www.mfy.org/wp-content/uploads/Disaster-Planning-Testimony-1-24-.

[28] National Council on Disability. Preserving our freedom ending institutionalization for people with disabilities during and after disasters, 2019. Available from: https://ncd.gov/publications/2019/preserving-our-freedom.

[29] Roth M, Kailes JI, Marshall M. Getting it wrong: An indictment with a blueprint for getting it right disability rights, obligations and responsibilities before, during and after disasters (Edition 1), 2018 May. Available from: http://www.disasterstrategies.org/index.php/news/partnership-releases-2017-2018-after-action-report.

[30] Department of Justice Civil Rights Division. Project Civic Access fact sheet. ADA.gov Department of Justice Civil Rights Division; 2021. [cited 2022 Jan 2]. Available from: https://www.ada.gov/civicfac.htm.

[31] Settlement Agreement between the United States of America and the City and County of Denver, Colorado under the Americans With Disabilities Act, DJ # 204-13-298 (2018). Available from: https://www.ada.gov/denver_pca/denver_sa.html.

[32] Settlement Agreement Between the United States of American and The City and County of Denver, Colorado Under the Americans with Disabilities Act DJ # 204-13-298 (2021). Available from: https://www.ada.gov/denver_pca/2021_denver_sa.html.

[33] Settlement Agreement Between the United States of American and The City and County of Killeen, Texas Under the Americans with Disabilities Act DJ 204-76-220. Available from https://www.ada.gov/killeen_tx_pca/killeen_sa.html.

[34] Medicaid. Home and Community Based Services 1915c [Internet]. Medicaid.gov. [cited 2022 Jan 13]. Available from: https://www.medicaid.gov/medicaid/home-community-based-services/home-community-based-services-authorities/home-community-based-services-1915c/index.html.

[35] State of Pennsylvania Department of Human Services. Department of Human Services Home and community based services general information. State of Pennsylvania Department of Human Services; 2022. [cited 2022 Jan 13]. Available from: https://www.dhs.pa.gov/Services/Assistance/Pages/Home-and-Community-Based%20Services.aspx.

[36] Taylor B. The development of emergency planning for people with disabilities through ADA litigation. J Marshall L Rev 2018;51:819.

[37] United States District Court, Central District of California. Communities Actively Living Independent and Free. City of Los Angeles, No. CV 09-0287 CBM (RZx); 2011. Available from https://dralegal.org/wp-content/uploads/2012/09/order_0.pdf.

[38] Complaint, California Foundation for Independent Living Centers v. City of Oakland, RG07339865 (Cal. Super. Ct. Aug. 9, 2007)., 2007. Available from: https://dralegal.org/case/california-foundation-for-independent-living-centers-cfilc-et-al-v-city-of-oakland-et-al/#files.

[39] United Spinal Association v. District of Columbia, et al Civil Action No. 14-1528 (CKK) Settlement Agreement., 2019. Available from: https://dralegal.org/press/disability-advocates-and-district-of-columbia-reach-landmark-settlement-to-protect-the-lives-of-people-with-disabilities-when-disaster-strikes/.

[40] Settlement Agreement. California Foundation for Independent Living Centers (CFILC), et al. v. City of Oakland, et al, 2009. Available from: https://dralegal.org/case/california-foundation-for-independent-living-centers-cfilc-et-al-v-city-of-oakland-et-al/.

[41] Norbert Enos, et al. v. State of Arizona, et al. No.CV_16-00384-PHX-JJT. (D. Ariz. February 10, 2017)., 2017. Available from: https://casetext.com/case/enos-v-arizona.

[42] Case Resolution Stipulation (Re: State of Arizona), Enos v. State of Arizona, No. 2:16-cv-00384-JJT (July 3, 2018). Cited in Taylor B. The development of emergency planning for people with disabilities through ADA litigation. J. Marshall L. Rev. 2017;51:819.

[43] Loye v. County of Dakota 625 F.3d 494 (8th Cir.2010)., 2010. Available from: https://www.courtlistener.com/opinion/179342/loye-v-county-of-dakota/.

[44] Disability Rights Education & Defense Fund. Preventing discrimination in the treatment of COVID-19 patients: The illegality of medical rationing on the basis of disability, 2020 Mar. Available from: https://dredf.org/the-illegality-of-medical-rationing-on-the-basis-of-disability/.

[45] Hodge J. Legal & regulatory challenges to crisis standards of care (CSC) in response to COVID-19. The Network for Public Health Law; 2020. [cited 2022 Jan 2]. Available from: https://www.networkforphl.org/resources/legal-regulatory-challenges-to-crisis-standards-of-care-csc-in-response-to-covid-19/.

[46] U.S. Department of Health and Human Services Office for Civil Rights. OCR reaches early case resolution with Alabama after it removes discriminatory ventilator triaging guidelines. U.S. Department of Health and Human Services Guidance Portal; 2020. [cited 2022 Jan 2]. Available from: https://www.hhs.gov/guidance/document/ocr-reaches-early-case-resolution-alabama-after-it-removes-discriminatory-ventilator.

[47] Robert T. Stafford Disaster Relief and Emergency Assistance Act, Public Law 93–288, as amended, 42 U.S.C. 5121 et seq. in Stafford Act, as Amended, and Related Authorities FEMA P-592. 05/2019, 2019. Available from: https://www.fema.gov/sites/default/files/2020-03/stafford-act_2019.pdf.

[48] Patient Protection and Affordable Care Act of 2010, 42 USC 18116, Section 1557, Available from: https://www.law.cornell.edu/uscode/text/42/18116.

[49] The Civil Rights Act of 1964, Title VI, 42 U.S.C. § 2000d et seq, as amended.

[50] U.S. Departments of Justice, Homeland Security, Housing and Urban Development, Health and Human Services, and Transportation. Guidance to state and local governments and other federally assisted recipients engaged in emergency preparedness, response, mitigation, and recovery activities on compliance with Title VI of the Civil Rights Act of 1964, 2016. Available from: https://www.justice.gov/crt/fcs/EmergenciesGuidance.

[51] Orenstein DG. When law is not law: setting aside legal provisions during declared emergencies. J Law Med Ethics 2013;41:73–6.

[52] Botoseneanu A, Wu H, Wasserman J, Jacobson PD. Achieving public health legal preparedness: how dissonant views on public health law threaten emergency preparedness and response. J Public Health 2011;33(3):361–8.

[53] Centers for Disease Control and Prevention Health. Insurance Portability and Accountability Act of 1996 (HIPAA). Centers for Disease Control and Prevention; 2018. [cited 2022 Jan 2]. Available from: https://www.cdc.gov/phlp/publications/topic/hipaa.html.

[54] The U.S. Department of Health and Human Services Office of the Assistant Secretary for Preparedness and Response, the Technical Resources, Assistance Center, and Information Exchange (TRACIE). HIPAA and disasters: What emergency professionals need to know, 2017 Sep. Available from: https://files.asprtracie.hhs.gov/documents/aspr-tracie-hipaa-emergency-fact-sheet.pdf.

[55] U.S. Department of Health and Human Services. March 2020 COVID-19 & HIPAA Bulletin limited waiver of HIPAA sanctions and penalties during a nationwide public health emergency, 2020 Mar. Available from: https://www.hhs.gov/sites/default/files/hipaa-and-covid-19-limited-hipaa-waiver-bulletin-508.pdf.

[56] U.S. Department of Health and Human Services. Is the HIPAA Privacy Rule suspended during a national or public health emergency? HHS.Gov Health information privacy; 2013. [cited 2022 Jan 2]. Available from: https://www.hhs.gov/hipaa/for-professionals/faq/1068/is-hipaa-suspended-during-a-national-or-public-health-emergency/index.html.

[57] Centers for Medicare & Medicaid Services. Coronavirus waivers & flexibilities. Centers for Medicare & Medicaid Services; 2021. [cited 2022 Jan 2]. Available from: https://www.cms.gov/about-cms/emergency-preparedness-response-operations/current-emergencies/coronavirus-waivers.

[58] U.S. Department of Justice Civil Rights Division. Civil rights and COVID-19, 2021 May. Available from: https://www.justice.gov/crt/Civil_Rights_and_COVID-19.

[59] U.S. Department of Homeland Security FEMA. A whole community approach to emergency management: Principles, themes, and pathways for action FDOC 104-008-1, 2011 Dec. Available from: https://www.fema.gov/sites/default/files/2020-07/whole_community_dec2011__2.pdf.

[60] U.S. Department of Health and Human Services Office of the Assistant Secretary for Preparedness and Response. 2017-2022 Health care preparedness and response capabilities, 2016. Available from: https://www.phe.gov/preparedness/planning/hpp/reports/documents/2017-2022-healthcare-pr-capablities.pdf.

[61] Centers for Disease Control and Prevention. Public health emergency preparedness and response capabilities, 2018. Updated 2019. Available from: https://www.cdc.gov/cpr/readiness/00_docs/.

Whole community as inclusive emergency management and public health preparedness

8

Susan Wolf-Fordham[a,b]

[a]Consultant, Association of University Centers on Disabilities, Silver Spring, MD, United States, [b]Adjunct Faculty of Public Health, Massachusetts College of Pharmacy and Health Sciences, Boston, MA, United States

Learning objectives

- Explain the "whole community" and "access and functional needs" concepts and the "CMIST" framework and how the concepts and framework can be implemented as part of public health disaster preparedness.

- Describe Community Stakeholder Meetings to review local emergency plans for strengths and gaps related to disability inclusion. Describe which stakeholders (groups and individuals) should participate in a Community Stakeholder Meeting.

- Explain how civic engagement relates to whole community emergency planning and response.

Introduction

The idea of the inclusion of people with disabilities into disaster management and public health preparedness continues to mature and evolve. There is no established "correct" way to accomplish it in practice. Seeking to fulfill its role of providing thought leadership and practical guidance, the United States (US) federal government has adopted the "whole community" philosophy or approach as a mindset that can be used by federal, state, territorial, tribal, and local government entities in developing and implementing public policy in collaboration with communities and community members.

Integrating Mental Health and Disability Into Public Health Disaster Preparedness and Response
https://doi.org/10.1016/B978-0-12-814009-3.00011-8

In contrast to the traditional top down approach, whole community leverages the strengths and resources of community members, businesses, non-profit organizations, schools, and other community assets, in addition to those of the government. While the traditional approach to emergency management and public health preparedness has historically excluded the needs and strengths of potentially vulnerable or marginalized populations, such as children, older Americans, people with limited English language skills or financial resources, individuals with disabilities, chronic and mental health conditions, and other populations, the whole community approach deliberately includes them.

These traditionally excluded groups are considered populations with "access and functional needs" (AFN). Having AFN may impact day to day life, and may also interfere with the ability to access or receive appropriate emergency services before, during, or after a disaster. As part of a local civic engagement process, inclusive practices lead to stakeholder participation in disaster-related activities to improve people's lives and community resilience. For many years scholars have studied civic engagement in multiple public activities. These studies establish an evidence-base for the effective use of civic engagement. One example of an inclusive whole community approach in the emergency management and public health preparedness fields is the Community Stakeholder Meeting (CSM) model tested in some Massachusetts (MA) communities. In this model, CSMs are collaborative community meetings to assess disability inclusion-related strengths and gaps in local emergency plans. This kind of approach is now being shared throughout the United States

While "whole community" and "access and functional needs" concepts apply to emergency management and public health preparedness at the local, state and federal levels, focusing on the local community level is a useful point of departure. Former Speaker of the US House of Representatives, Thomas P. O'Neill Jr. of MA often said, "all politics is local" [1]. There is also a saying in the emergency field that all response is local because emergency response is initially implemented at the local level [2]. The local level is the one closest to the community, to the people who will be directly impacted by disasters, and is where Americans have a strong, direct influence on their government.

The whole community philosophy and approach

It is in this context that FEMA developed its whole community philosophy to emergency management. As defined by FEMA, whole community

> ...is a means by which residents, emergency management practitioners, organizational and community leaders, and government officials can collectively understand and assess the needs of their respective communities and determine the best ways to organize and strengthen their assets, capacities, and interests. By doing so, a more effective path to societal security and resilience is built [3, p. 3].

In other words, the whole community emergency approach centers on engaging all of the community in each of the four emergency management phases to ensure that emergency mitigation, planning, response and recovery address the needs of the whole community.

FEMA supported the whole community philosophy in response to criticisms the agency received asserting poor responses to Hurricane Katrina in 2005 and to other disasters. Other factors driving the approach include an increased awareness of the needs of potentially vulnerable populations, the increasingly high cost of emergencies with ongoing budget constraints, and the realization that governments couldn't handle these events alone [3,4]. Public engagement builds public trust and engenders legitimacy for policies developed based on public opinion [5]. FEMA incorporated the whole community approach into its major guidance documents such as the National Preparedness Goal [6]. FEMA'S Comprehensive Preparedness Guide (CPG 101) [7], the basis for US public safety emergency planning, states that community engagement is "crucial for the success of" emergency operations plans [7, p. 46] and outlines several steps to take toward whole community planning. However, CPG 101 also cautions that "[d]etermining how to engage the community effectively ... is one of the biggest challenges that planners face [7, p. 46]", and argues that misunderstanding about the public interest, safety, or limited planning may be the cause.

On a broader level, the whole community approach is supported by policies and guidance from a number of federal agencies (see Table 1). These guidance documents stress the importance of community engagement, leveraging community resources, and conducting inclusive planning with and for people with disabilities and others with access and functional needs.

Despite limited evaluation, FEMA and others report that whole community approaches result in benefits such as creating relationships to strengthen social bonds, enhanced preparedness and response, and strengthening local and national resilience [3,14]. Other reported benefits include greater knowledge of a community's complexity, capabilities, and needs, all of which empower local action and fosters relationships with community leaders. Horney and colleagues state that participation in emergency planning can improve emergency plans and the act of planning can further engage residents [15]. The authors examined local hazard mitigation plans, a usual component of a complete community emergency plan, and found that when planners supported community member engagement in hazard mitigation, the resulting plans had 76% more mitigation measures than comparison plans developed without engagement. Including people with disabilities in the planning process benefits the large number of people with disabilities among the population and enriches emergency planning for the entire community [16,17].

Finally, the whole community approach comports with the Americans with Disabilities Act (ADA) [18,19] and Section 504 of the Rehabilitation Act of 1973 [20], which prohibit disability discrimination. The approach also is in keeping with civil rights laws such as the Civil Rights Act of 1964, which prohibits discrimination based on color, religion, or national origin [21], and the Affordable Care Act's

---

Table 1 Documents reflecting US agencies using the whole community approach.

- Center for Disease Control and Prevention (CDC), Public Health Preparedness Capabilities [8]
- FEMA, Comprehensive Preparedness Guide (CPG 101) [7]
- FEMA, National Response Framework [9]
- FEMA, National Preparedness Goal [6]
- US Department of Education, Readiness and Emergency Management for Schools Readiness and Emergency Management for Schools (REMS) articles on the REMS website [10]
- US Department of Health and Human Services Office of the Assistant Secretary for Preparedness and Response (ASPR), Health Care Preparedness and Response Capabilities [11]
- US Department of Health and Human Services Office of the Assistant Secretary for Preparedness and Response (ASPR),US National Health Security Strategy and Implementation Plan (NHSSIP) [12,13]

anti-discrimination provision designed to reduce disparities and improve outcomes for people with disabilities and many other historically marginalized populations [22].

Barriers to the whole community approach have been reported. The US National Health Security Strategy (NHSS) and Evaluation Plan 2015–2018 notes the ongoing challenge of engaging communities such as populations with access and functional needs [23]. The NHSS Implementation Plan 2019–2022 addresses this gap by broadening the concept of public health and healthcare to include behavioral health and social services and adds an action step to create partnerships with community based organizations [24]. Other reported challenges include potentially unrealistic community member expectations, limited funding, challenges evaluating whole community efforts, and sustaining long term collaborator interest [25]. Knowledge gaps about the disaster-related needs of people with AFN can also be barriers. Several authors have written about emergency manager misconceptions about the emergency needs of, and legal requirements for, people with disabilities [16,26]. Other authors document similar local public health preparedness planner knowledge gaps [27,28].

Whole community philosophy statements in local emergency plans

Despite these barriers, some local emergency plans include statements supporting the whole community approach. Examples are shown in Fig. 1 below.

Note that while each plan contains different wording about the whole community philosophy they have common features. Both plans list key whole community stakeholder groups. The Miami-Dade County plan also notes the stakeholder engagement methods used, including committee meetings throughout the year and consultation about disability-related issues. The Miami-Dade County plan refers to people with disabilities specifically while the Monterey County plan does not. The Monterey County plan, on the other hand, focuses on building local capacity and creating resilience via community engagement.

Some jurisdictions have taken the additional step of passing state laws and local ordinances or directives that mandate whole community planning. For example, Los Angeles Executive Directive No. 23 (2012) is an example of a mayor's executive order that requires all city departments to create emergency plans using a whole community approach [31]. California Assembly Bill No. 2311 (2017) is a state statute that requires each county to integrate populations with access and functional needs into city and county emergency plans [32].

Individuals with access and functional needs

The whole community philosophy includes ensuring that diverse community members and organizations representing them have a seat at the local emergency planning table and that their needs are addressed in emergency plans as well as in actual practice. The question arises as to which populations need considerations in planning and response that go beyond that of the general population. Before and after Hurricane Katrina in 2005, individuals with disabilities and many other people were considered to be populations with "special needs," and it was recommended that emergency and public health preparedness planners take this population's needs into account in emergency planning. However, the definition of "special needs" was very broad.

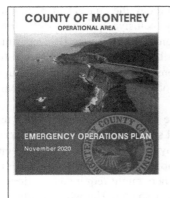	Monterey County, California, Emergency Operations Plan, 2020 Draft, under review by the Board of Supervisors "Monterey County has embraced FEMA's whole community approach to creating engaged and resilient communities by which residents, emergency management practitioners, community leaders and government officials can understand and assess the needs of their respective communities and determine the best ways to organize and strengthen their assets, capabilities, and interests. By engaging communities, we can understand the unique and diverse needs of a population…. Local capacity is built on the empowering of community members, social and service groups, faith-based and disability groups, academia, professional, private and nonprofit sectors to strengthen what works in their communities on a daily basis. Existing structures and support organizations can be leveraged and empowered to act during and after a disaster strikes." [29]
	Miami-Dade County, Florida, Miami-Dade County, Florida Comprehensive Emergency Management Plan (CEMP), 2017 "Throughout the CEMP volumes, the Federal Emergency Management Agency (FEMA) Comprehensive Preparedness Guide (CPG) 101 fundamentals and the Whole Community approach strategy are utilized. While CPG 101 provides emergency management professionals with guidance on plan development, the Whole Community approach refers to the collaboration between emergency management and a wide range of county stakeholders to ensure that all are part of the planning process. These stakeholders include residents, vulnerable populations, organizational and community leaders, faith-based and non-profit organizations, the private sector, and government officials. These groups are regularly engaged, and are leveraged in various committee meetings throughout the year. Great care has been taken to consult with disability organizations as well as the Americans with Disabilities Act of 1990 as amended and the Guidance on Planning for Functional Needs Support Services in General Population Shelters during the plan development processes." [30]

FIG. 1

Examples of local emergency plans with whole community philosophy statements [29,30].

> Special needs populations…include an extremely broad group of people, including people with disabilities, people with serious mental illness, minority groups, non–English speakers, children, and the elderly. Other lists add single working parents, people without vehicles, people with special dietary needs, pregnant women, prisoners, people who are homeless, and others [33].

Advocates such as June Isaacson Kailes and Alexandra Enders [33] argued that this terminology was too broad and imprecise to operationalize, leading to inefficient disaster plans, which in turn led to many failures of disaster response. They maintained that considering the population's functional limitations and related needs should be the focus of more specific emergency planning and more tailored response activities. Kailes later said that "'It's not the *diagnosis* but the *needs* that should be addressed'" (emphasis in original) [34].

The idea of populations with functional needs has evolved and today the concept of populations with "access and functional needs" is part of the whole community inclusive approach. Both FEMA and the Office of the Assistant Secretary for Preparedness and Response (ASPR) at the US Department of Health and Human Services have definitions of "access and functional needs." The FEMA definition focuses on functional needs [35], while the ASPR definition considers both access and functional needs [36]. However, there is general consensus about the meaning of the terms and the populations to which the phrase applies.

Populations with access and functional needs encompass people with "*access needs*" which refer to access to services, information and other resources, and physical space [36]. Someone with access needs might need a ramp for physical access to a building or other physical space such as an emergency shelter. Someone else with access needs might require modifications for communication access to shelter registration forms or emergency warnings and alerts, and others might require adaptations for equal access to programs, services and activities offered in a disaster shelter. The whole community approach also includes people with *functional needs* [36], limitations for which someone may need support, modifications, or accommodations. For example, if someone cannot hear disaster warnings via a bullhorn (a functional hearing limitation), then they will need to receive the warning via another modality, such as text. Or, if someone needs support for daily living activities such as eating or bathing (functional self-care limitations), they will need the same support in an emergency shelter. As well, someone else may have a functional cognitive limitation that would necessitate support to understand how to follow personal preparedness instructions. Of course, someone may have more than one access and/or functional need and a functional need may lead to an access need or vice versa. The whole community philosophy is not about labels or diagnoses. Two people with the same diagnosis may have very different access and functional needs. For example, one person with an autism diagnosis or label may be extremely self-sufficient, live on their own, and work independently, while another person with the same diagnosis may require 24-7 care and support. Fig. 2 provides examples of the kinds of populations with likely access and functional needs.

The CMIST framework

In order to better address the concept of access and functional needs, the CMIST framework was developed as a way to integrate potential functional needs into a useful conceptualization for emergency planning and response purposes. The framework provides a flexible, cross-cutting approach without regard to status, label or specific diagnosis [33,37]. As seen in Fig. 3, this framework contains five broad categories of functional need: communication, maintaining health, independence, self-determination, support and safety, and transportation. The acronym CMIST, pronounced "see mist," is a mnemonic for the first letters of each category.

CMIST can be used for temporary and permanent conditions and the categories are cross cutting. For example, a person with a broken leg and someone without a car may both have transportation needs and benefit from an assessment of their CMIST needs. Communication needs can be as diverse as the use of screen reader software for a blind person to read text on a computer screen and the need for evacuation instructions to be written in plain language or with pictures for someone with an intellectual disability. Individuals may also have more than one CMIST need. For example, someone who takes insulin for diabetes and uses a wheelchair for mobility has two CMIST needs, maintaining health and

FIG. 2

Examples of populations with likely access and functional needs.

Credit: National Center on Disabilities in Public Health. Association of University Centers on Disabilities (AUCD). Prepared4ALL whole community inclusive emergency planning. Association of University Centers on Disabilities (AUCD); 2021 [cited 2021 Oct 21]. Available from: https://nationalcenterdph.org/our-focus-areas/emergency-preparedness/prepared4all/online-training/. Used with permission.

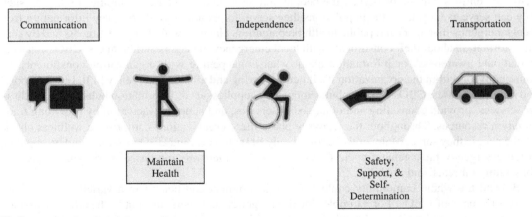

FIG. 3

The CMIST framework.

Credit: Adapted from United States Department of Health and Human Services Office of the Assistant Secretary for Preparedness and Response. At-risk individuals. Public Health Emergency. U.S. Department of Health and Human Services Office of the Assistant Secretary for Preparedness and Response; 2021 [cited 2021 Dec 21]. Available from: https://www.phe.gov/Preparedness/planning/abc/Pages/atrisk.aspx.

independence. The point of the CMIST framework is not to demarcate a particular group or population. Instead, the access and functional needs concept with its CMIST framework stands as a reminder to emergency and public health preparedness planners and practitioners of the major domains of need that should be evaluated, planned for, and addressed for individuals and populations with disabilities and others with access and functional needs.

The important role of disability and other community-based organizations within the whole community approach

The chapter now turns from populations with access and functional needs to other key actors in the whole community approach. The implementation of whole community requires purposeful outreach to and engagement with stakeholders in each community. Community-based organizations (CBOs) can play an important role in the whole community approach. These local entities, often not-for-profit organizations, include aging, disability and child welfare organizations, social services organizations, advocacy organizations, area agencies on aging and disability-led independent living centers, among others. Faith-based organizations have been found to be particularly important partners in both disaster planning and response [38]. Stakeholders from these organizations may include service recipients or program participants, members, volunteers, donors, board members and staff, and those who attend organization events. Local public health departments commonly collaborate with community organizations [14,39] and FEMA strongly promotes this kind of collaboration by providing numerous resources, FEMA liaison staff and a detailed CBO online training (see, e.g., Refs. [40,41]).

Community-based organizations remain natural and critical bridges between local governments and AFN populations. Whether disability focused or organized around other affiliations or affinities, CBOs are entities that are trusted by the people they serve and by their communities [3,42]. As such, they can serve as liaisons to the populations they serve, for example by sharing key information from local emergency managers and public health preparedness planners with their constituents and by sharing information about their constituents with local emergency managers and planners. CBOs can share "situational awareness" or information about what is happening with organization constituents, the community, or within the organization itself before, during and after an emergency [43]. In the response and recovery phases, CBOs can distribute emergency supplies, act as information hubs for initial disaster services, provide counseling and other social services, and support evacuation by providing transportation resources. Throughout the recovery phase these organizations can provide wellness checks for the people they serve. Many of these roles likely fit within existing CBO missions, and by engaging in the emergency management cycle CBOs can expand their mission-based work while increasing community outreach and visibility.

Beyond the whole community context, local government can benefit in a variety of ways from the involvement of CBOs. For example, local emergency managers and public health preparedness planners may expand their own knowledge and capabilities and have their reach extended by leveraging CBO capabilities to meet community needs. CBOs can also be a source of untapped resources that could benefit the community while decreasing cost, personnel or other burdens faced by local emergency management and public health departments. "[T]hese communities possess key cultural assets that may not be available to official emergency responders (e.g., bi-lingual/bi-cultural community outreach workers) and key physical assets that could be activated in disaster (e.g., industrial kitchens in community centers)." [44, p. 5]. While the importance of CBO engagement in whole community activities is clear, successful whole community activities are predicated on CBO, public health and emergency management personnel beliefs about the whole community philosophy.

Emergency manager and public health personnel beliefs about the whole community approach

In an effort to assess the potential for government emergency and public health preparedness planners to leverage CBO capabilities, researchers have examined CBO attitudes toward whole community planning. Recent findings sound a note of caution for future efforts. Multiple barriers to CBO engagement with local emergency and public health departments exist, including lack of ongoing mutual relationships, misalignment of goals, and community organization financial and personnel resource gaps [44]. Goal alignment is particularly important for success [45].

Some research indicates that emergency managers see whole community planning as a positive opportunity to engage the public [46]. However, others argue that while emergency managers report valuing civic engagement, they actually engage in emergency planning only with consultants or a small local government planning team [47] and few emergency managers make targeted outreach to potentially vulnerable populations [15].

Similar findings have been reported in the public health domain, which has been more extensively studied. It appears that while local public health agencies have deepened whole community engagement over time, stagnation due to budgetary and staff constraints has been reported [48]. In addition, public health preparedness planners have reported concerns that residents will have unrealistic expectations of the community's emergency capabilities; these concerns could bias public health preparedness planner motivation to collaborate [14]. Like emergency planners, public health preparedness planners may face knowledge gaps (see, e.g., Ref. [28]). There is evidence that many local health departments lack an understanding of disability demographics and the concept of disability as a health disparity population. The evidence also indicates a limited understanding of successful approaches to deepen collaborative relationships with people with disabilities and disability organizations (see, e.g., Refs. [28,39,49,50]). In the aggregate, these findings raise substantial concerns about whether local governments currently possess sufficient knowledge, interest, and capacity to successfully implement a whole community program.

Learning from civic engagement literature and practices

The whole community approach is built on the foundation of incorporating stakeholder input into emergency planning and response. As noted above, obtaining this input has not been straightforward. While the term "civic engagement" is generally not used in the whole community literature, there is the potential that the whole community field can benefit from evidence-based civic engagement practices. It is also possible to view whole community as a form of civic engagement relate to emergencies and disasters.

There is no standard definition for civic engagement. One workable civic engagement definition is "Public participation in governance [that] involves the direct involvement – or indirect involvement through representatives – of concerned stakeholders [persons, groups, or organizations] in decision-making about policies, plans or programs in which they have an interest." [51, p. 158]. This is a broad, neutral definition that seems to apply to the whole community approach, with community member involvement with policies, programs and plans related to emergencies, pandemics and disasters.

A conceptual framework for civic engagement

There are a variety of conceptual frameworks to describe the interactions between community members and governments based on depth of participation and implementation methods. These frameworks are useful for considering the depth of the public's engagement in a civic activity, like whole community activities. One illustrative example is the "Spectrum of Public Participation" [52].

Developed and licensed by the International Association for Public Participation (IAP2) since 1990, the Spectrum of Public Participation frames the decision making relationship between public decision-makers and their affected stakeholders, and offers a transparent set of engagement choices. The framework is grounded in IAP2 Core Values and its Code of Ethics [53]. The objective is to have decision makers purposefully choose one of five levels of stakeholder engagement in public decision making and then make commitments to stakeholders about whether and how their inputs will be used. These five levels include: (1) informing the public about an issue; (2) enabling public consultation to provide feedback; (3) incorporating public opinion into public decisions in some way; (4) collaboration where public opinion is incorporated into decisions to the fullest possible extent; and (5) empowering the public to make the final decision [52]. Frameworks such as this can be used to inform, categorize, and evaluate civic engagement with disaster management and public health agencies and practitioners.

Civic engagement impacts

While there is much civic engagement literature, there is limited systematic research evaluating civic engagement efforts. Public officials have an awareness of civic engagement, but also have varied opinions about its effectiveness [54]. There is some evidence that direct participatory civic engagement can affect local policy and emerging evidence about the utility of online engagement. An important related issue is what to measure and how to define the success of a public engagement process [55].

Most studies indicate positive results for stakeholders, including knowledge and skill gains [54]. Civic engagement activities in which community members have greater direct input into policy lead to more positive outcomes than public hearings and online methods. There is also some evidence that the more deliberative and participatory kinds of civic engagement activities can enhance feelings of community responsibility and favorability toward government-citizen collaborations, deepen determination to participate, and generate new ideas [54].

Researchers [56] applied a civic engagement lens to public participation in recovery from a flood disaster in two adjacent communities. They state that while administrators agree about the importance of civic engagement in theory, there is disagreement about how civic engagement should look in practice. They note that communities with more citizen-engaged processes had more successful flood recovery as expressed in post-flood recovery election results and engagement process satisfaction data. Processes contributing to this success included community members receiving clear information, having their questions addressed, and participating in decision making, most of which did not happen in the less successful community. Nonetheless, the researchers report that perception of participation and impact, rather than actual participation, was the best indicator of community member satisfaction and trust in a civic engagement process [55]. The authors conclude that for community members "participation has a symbolic benefit that may be more important than the instrumental role" [55, p. 420].

The whole community approach in practice: Community Stakeholder Meetings

As noted above, deciding how to engage the community in emergency-related activities can be a challenge [7]. Varied sources describe potential and current examples of local public health department and emergency management agency participatory and inclusive initiatives. Examples of these activities include conducting surveys or trainings for the entire community, publishing emergency plans for public comment, holding community forums, creating public health emergency preparedness materials in non-English languages, and collaborating with outside organizations to leverage resources [57,58]. People with disabilities can participate in relevant training programs or exercises, and public health preparedness planners and emergency managers can make outreach to disability service providers and community disability groups, as well as seeking input about plans and policies from people with disabilities [16].

A real world example of the whole community approach in practice is the Community Stakeholder Meeting (CSM) process developed at the Eunice Kennedy Shriver Center, University of Massachusetts Chan Medical School [59]. The Shriver Center, a Universitry Center of Excellence in Developmental Disabilities (UCEDD), is dedicated to education, research, and community service to improve the quality of life for children and adults with intellectual and developmental disabilities and their families [60]. The CSM process was originally part of a project developed in 2012–2013, initially funded by a FEMA grant through the Massachusetts Executive Office of Public Safety and Security [59]. Under different funding after 2014, the concept continued and evolved. In 2020, the Community Stakeholder Meeting concept was incorporated into the Prepared4ALL initiative, Association of University Centers on Disabilities (AUCD) [43]. Two online course modules [43,Lessons 7 and 8] explain access and functional needs, CMIST, and the whole community philosophy, including an explanation of Community Stakeholder Meetings, and are accompanied by a related workbook guide to lead communities through the CSM process [61].

Community Stakeholder Meetings involve a collaborative, inclusive community engagement approach to assess the level of disability inclusion in local emergency plans [59]. A CSM is a whole community emergency planning tool that uses a workbook guide developed for this purpose. In the process, local community members with disabilities, families, allies and support networks, disability organizations and service providers, emergency management and public health preparedness planners, first responders and local government officials come together to collaboratively review the local emergency plan(s) and community response practices with a goal of identifying strengths and gaps related to disability inclusion. CSM participants discuss response practices, as well as reviewing plan(s), because sometimes local plans do not capture what may actually happen in practice. Meeting participants brainstorm strategies and develop an action plan to build on emergency plan strengths. While CSMs take time and persistence to plan and coordinate, the benefits may be great. In addition to closing gaps related to disability inclusion as required for meeting whole community needs and legal compliance, CSMs engage in information and resource sharing, raising awareness, and building new and strengthening ongoing relationships. CSMs may also encourage community members with disabilities to become more active locally in other endeavors. The long term goal of all of these interactions is to strengthen community resilience [59].

As CSM participants work together to complete the Active Planning Workbook, group discussion will identify areas of strength and gaps, and lead to strategies for closing the gaps based on those strengths. Once the Workbook is complete the community will have a blueprint to make the local emergency plan (and any practices not in the plan) more inclusive and a time frame in which to do so [59]. Fig. 4 summarizes the Workbook's table of contents, and indicates the order of the discussion at a CSM. Each part of the meeting has a Workbook tool associated with it.

The majority of CSM meeting time is spent reviewing the topical checklists that comprise Tool #1, including a local needs assessment, demographics, and other checklists. The meeting facilitator reads the Tool #1 checklists aloud, item by item. The checklist items represent standards. CSM participants decide together whether the plan and practices meet each standard. The checklist is meant to start a discussion about people with disabilities, and their needs, relationship with and roles in local emergency and public service delivery, community resource availability, whether the plan addresses access and functional needs, and individual responsibility to self-prepare. During the discussion of each item, participants share ideas, knowledge, expertise, and resources [59].

The Active Planning Workbook Topic Outline

The Community Stakeholder Meeting (CSM) process

Whole community emergency planning
      a. Functional and access needs and the Communication-Maintain Health-Independence-Support/Safety/Self-
         Determination-Transportation (C-MIST) framework

Tool 1, Where We Are Now? Gap Analysis/Needs Assessment Checklists
      a. Local needs assessment and demographics
      b. Local emergency planning process
      c. Public emergency preparedness education
      d. Emergency communication (e.g., warnings, alerts)
      e. Transportation and evacuation
      f. Mass care shelters (including shelter set-up, operation, and deactivation)
      g. Hazardous materials decontamination
      h. Emergency dispensing sites (including COVID-19 vaccine sites)
      i. Recovery

Tool 2, Where Do We Want to Be? Tool to set priorities.

Tool 3, How Do We Get There? Closing the Gaps with an Action Plan Tool to create an action plan.

FIG. 4

The Active Planning Workbook topic outline [61].

Credit: Adapted from Wolf-Fordham S, Shea N, Gleason P, Hamad CD. Revised by Wolf-Fordham S. Active planning workbook meeting the emergency & disaster needs of people with disabilities; 2015, Revised 2021. Available from: https://nationalcenterdph.org/wp-content/uploads/2021/07/Emergency-Planning.pdf.

Frequently, during CSMs there is disagreement about whether a particular standard has been met, and this is an important touchpoint because it highlights knowledge and communication gaps among meeting participants and sometimes between people working in the same local government agency. The CSM process requires that disagreements be resolved by clarifying whether and how a standard may have been met in whole or in part before the group can move onto the next checklist item. As the discussion continues, gaps and areas for improvement become clear and participants begin to think of strategies to address the gaps [59].

After all the checklists have been reviewed, the CSM meeting participants decide on five gaps or areas for improvement that have simple, quick and/or inexpensive solutions. Addressing these issues will provide early successes and hopefully spur the group to tackle more complex gaps. Participants then choose gaps that will need more extensive or expensive solutions and likely require more time to develop and implement remediation strategies. The final step of the process is development of an Action Plan describing the main gaps identified, proposed strategies to address those gaps, the people responsible for implementation, and a timeline for addressing the gaps. The Action Plan is based on the results of the checklist and discussions at the CSM [61].

At some point during most CSM meetings, there is an "Aha!" moment that helps move the group from identifying gaps to gap closing strategies, often before formally setting priorities or creating the Action Plan. For example, at one CSM, participants discussed the kind of equipment to store for emergency use. The discussion turned to the likely need for wheelchairs during emergency response in case survivors were injured or had lost personal mobility equipment. While the group agreed that wheelchairs were necessary, the local budget was limited and there was concern about the items' cost. The local government's Council on Aging director attending the meeting shared that her office had extra wheelchairs and offered them for community use during an emergency, including storing them until they were needed. This kind of discussion, in many different forms, occurred at every CSM held in MA [43,Lesson 8].

During CSMs held between 2012 and 2014, all 25 participating localities identified at least 5 emergency plan gaps or areas for improvement related to disabilities in their local emergency plan and/or response practice. A review of completed workbooks from 11 of the 19 FEMA funded CSMs indicated common themes among the gaps, as described in Table 2 below. Common gap areas identified by the participating communities included communication, sheltering, evacuation, transportation, and a lack of knowledge of the local disability community, findings supported by the literature (see, e.g., Refs. [27,28]). All of the CSMs developed mainly low cost (and thus within municipal budgets) strategies to close the gaps, demonstrating feasibility and utility of the process [59]. Representative remediation strategies included changes to sheltering and evacuation plans, improving communication and information sharing, using mapping technology, and identifying new sources for resources. Table 3 provides an example of representative gap closing solutions identified via the project.

CSMs present an opportunity for local emergency management, public health organizations, and healthcare coalitions to engage community partners in preparedness efforts, assess and enhance public health and emergency plans, and access potential new resources. Public health and emergency management organizations have different priorities and organizational cultures, often impeding partnership [63], however a CSM offers the opportunity to bridge that gap. The meeting provides the circumstance for local public health, emergency management and other local government entities, for profit and non-profit organizations, and community groups and their members to collaborate. The meetings also offer emergency managers, public health preparedness planners, responders, and disability community members the chance to learn more about each other while working toward a common goal.

Table 2 Representative gaps identified at 11 community stakeholder meetings held in 2012–2013 [62].

Gaps	# Communities (*N*=11)[a]
Limited knowledge about disability community needs	6 (55%)
• No disability demographics	
• Disability demographics tracked by multiple local government departments but no sharing	
• No system to update disability demographics	
• Poor communication with Deaf community	
Communication: Public education, alerts and warnings, communication with potentially vulnerable groups	11 (100%)
• Lack of American sign language/non-English language interpretation	
• Lack of information in alternate formats	
• Lack of emergency-related information on local government website	
• Local government website not accessible	
• Confusion about accessibility of local reverse 9-1-1 system	
• Reverse 9-1-1 system doesn't support adaptive communication technology	
Planning	10 (91%)
• No Memoranda of Understanding (MOU) or no written MOUs formalizing relationships with partner organizations	
• Local emergency planning committee not inclusive	
• CERT/MRC volunteer response organizations don't include people with disabilities as members	
Mass care sheltering	7 (64%)
• Limited services	
• Lack of resources	
• Lack of procedures	
• Lack of knowledge about disabilities and reasonable modifications (accommodations)	
• Inadequate space	
Evacuation and transportation	2 (27%)
• Potentially competing local evacuation or transportation plans	
• Evacuation/transportation plan relies solely on the state for assistance	

[a] *The totals do not add to 100% because all communities had more than one gap or area of need.*

Table 3 Representative gap closing solutions identified at 11 community stakeholder meetings held in 2012–2013.

Gap	Solution
Limited knowledge of disability community	• Better assess local disability-related needs 　▪ Survey local disability organizations 　▪ Include disability organizations in planning 　▪ Conduct formal needs assessment 　▪ Use functional needs data from American Community Survey (US Census) to aid planning 　▪ Expand local census to include disability-related information 　▪ Add information about persons in congregate care settings to evacuation plan • Improve information sharing: Share disability demographic information among local government departments; resolve related privacy issues • Disability commission to host "think tank" of people with disabilities to discuss preparedness and response
Communication	• Enhance communication among groups 　▪ Local disability commission and emergency planners 　▪ Among local government departments 　▪ Among the local government, disability community, and non-English speakers 　▪ Hold meeting with community group homes and Independent Living Centers (disability-led community organizations) • Use communication conduits 　▪ Provide disability organizations with English and non-English information 　▪ Use disability organizations to disseminate public education and warnings to constituents • Develop alternate communication modalities/translations 　▪ Develop Braille information 　▪ Research Video Remote Interpreting (VRI) technology and specialized apps for direct communication with the Deaf community 　▪ Identify sign language interpreters and create resource list 　▪ Develop relationship with local churches to create list of on-call non-English language translators and community leaders 　▪ Enhance relationship with non-English speaking community 　▪ Post key emergency-related messages in local laundromats (this could include other places people may gather) • Review message content 　▪ Work with disability community and non-English speakers to ensure message clarity 　▪ Use word processing software readability tools to test messages for clarity before dissemination 　▪ Create easy to understand flow chart with emergency instructions • Review message effectiveness (e.g., with disability community)

(Continued)

Table 3 Representative gap closing solutions identified at 11 community stakeholder meetings held in 2012–2013—cont'd

Gap	Solution

Planning	• Increase number of planning partners/expand local planning committee
	  ▪ Continue holding meetings with the CSM group
	  ▪ Develop expanded local planning committee to include members of CSM group, disability community, housing authority, senior center, school system special education parent advisory council representatives in planning process and/or meetings
	• Make creative use of volunteers: Develop a list of local community members who would be willing to volunteer specific skills to the community in an emergency
	• Develop Memoranda of Understanding (MOU) re: transportation, shelter resources, and volunteers
	  ▪ Develop MOU with independent living centers and other disability and provider organizations
	• Include disability-related issues in post-disaster After Action Reports

Mass Care Sheltering	• Identify new accessible shelter sites • Identify and acquire resources: Borrow CCTV (video magnifier) from library for blind/low vision shelter residents; borrow Hoyer lift (for transfers of people with limited walking abilities) for shelter from local special education school; develop shelter resource list; develop plan to track shelter supplies for people with disabilities; identify supplemental oxygen sources; assess availability of additional generators • Develop policies/procedures regarding: Food allergies, need for extra space, privacy, service animals, personal assistance service providers, auxiliary power; medication security; medical/biohazard waste collection • Develop procedure for service animals in shelters; develop shelters that are service animal-friendly • Drills and training: Identify shelter worker roles and responsibilities; develop training for shelter managers regarding disability issues; incorporate disability issues into full scale shelter setup drills
Transportation	• Transportation services: Review contracts to understand bus company priorities-don't assume a particular community is a priority for the company because they likely have contracts for emergency transportation services with other communities; develop back up transportation contracts • Coordination with other transportation/evacuation plans • Reach out to local corporations to coordinate regarding their evacuation plans • Help local special education schools develop evacuation plans

Credit: Adapted from the Wolf-Fordham S, Shea N, Hamad CD, EK. Shriver Center University of Massachusetts Chan Medical School. Active planning for mass care sheltering and evacuation of people with disabilities final project report.

Conclusion

Public health preparedness and emergency management professionals face challenges planning for people with disabilities and other populations with access and functional needs. While the law requires that all federal, state, territorial, tribal, and local emergency and disaster plans address the likely needs of people with disabilities in all phases of a disaster [18–22], people with disabilities may still face barriers to accessing or receiving emergency-related services or participating in disaster-related activities before, during or after a disaster. As a planning philosophy and approach, thought leaders in the disability community, at FEMA, ASPR, and other federal agencies have adopted the whole community approach and the CMIST framework to identify and prepare for people with access and functional needs throughout the emergency life cycle. Participation of the whole community means equal access to local, state, territorial, tribal, and national emergency and pandemic-related activities and programs without discrimination. It means not only meeting the access and functional needs of all individuals, but also consistent and active participation in all aspects of planning and response by community members.

The choice of a participatory process to effect specific change creates both opportunities and challenges. From the literature it is clear that such a process can lead to reform and systems change, re-balance power, and modify citizen and administrator roles, and that incorporating inclusive participation processes can change outcomes. Connecting emergency management and public health with civic engagement strategies could enrich both fields with opportunities to expand learning, hopefully leading to better local emergency plans, and ultimately more lives saved.

On the other hand, the literature also makes clear that participatory processes are hard, complex and time-consuming and that they can be costly in terms of funding, opportunity, and the intellectual and emotional energy of the participants. Even if done well, participatory processes do not necessarily produce constructive, useful end-products. Experience is replete with examples of successful pilots and trailblazing programs that could not either be replicated or scaled. Decision makers need to assess the capabilities of their organizations and of their various stakeholders and then adopt planning processes that are well matched to the situation on the ground in their communities.

Questions for thought

(1) Reflect on the "access and functional needs" concept. Do you think any populations are missing from the definition? Has the COVID-19 pandemic highlighted any population(s) that should be added, and, if so, why should they be added?

(2) How might you apply the access and functional needs concept to increase inclusion and equity in public health outside of emergency management and public health preparedness planning?

(3) Think about holding a Community Stakeholder Meeting (CSM) to identify disability inclusion-related strengths and gaps in a local emergency plan. Make a list of the people (by role), resources, settings, and other elements that could make the meeting successful and a list of the people (by role), resources, settings, and other elements that could make the meeting challenging. Then make a third list with strategies to use the positives to overcome the negatives.

References

[1] Matthews C. Hardball: How politics is played told by one who knows the game. Simon and Schuster; 1999 Nov 2, ISBN:978-1-4165-6261-0.

[2] Congressional Research Service. Congressional primer on responding to and recovering from major disasters and emergencies. Update June 3 2020 [cited 2021 Dec 29]. Available from: https://sgp.fas.org/crs/homesec/R41981.pdf.

[3] U.S. Department of Homeland Security FEMA. A whole community approach to emergency management: Principles, themes, and pathways for action FDOC 104-008-1, 2011 Dec. Available from: https://www.fema.gov/sites/default/files/2020-07/whole_community_dec2011__2.pdf.

[4] Kapucu N. Emergency management: whole community approach. In: Dubnick M, Bearfield D, editors. Encyclopedia of public administration and public policy. 3rd ed. New York, NY: Taylor and Francis; 2015. p. 1–6.

[5] Schoch-Spana M, Chamberlain A, Franco C, Gross J, Lam C, Mulcahy A, Nuzzo JB, Toner E, Usenza C. Conference report disease, disaster, and democracy: the public's stake in health emergency planning. Biosecur Bioterror 2006;4(3):313–9. Available from: https://www.centerforhealthsecurity.org/our-work/events-archive/2006_disease_disaster_democracy/Conference%20Report.pdf.

[6] United States Department of Homeland Security. National preparedness goal. Washington, DC: United States Department of Homeland Security; September, 2015. Available from: https://www.fema.gov/sites/default/files/2020-06/national_preparedness_goal_2nd_edition.pdf.

[7] United States Department of Homeland Security FEMA. Developing and maintaining emergency operations plans comprehensive preparedness guide (CPG 101). Version 3. FEMA; 2021. Available from: https://www.fema.gov/sites/default/files/documents/fema_cpg-101-v3-developing-maintaining-eops.pdf.

[8] Centers for Disease Control and Prevention. Public health emergency preparedness and response capabilities, October, 2018. Updated January 2019. Available from: https://www.cdc.gov/cpr/readiness/00_docs/CDC_PreparednesResponseCapabilities_October2018_Final_508.pdf.

[9] United States Department of Homeland Security. National response framework. 4th ed. Washington, DC: United States Department of Homeland Security/FEMA; 2019. Available from: https://www.fema.gov/sites/default/files/2020-04/NRF_FINALApproved_2011028.pdf.

[10] United States Department of Education Readiness and Emergency Management for Schools REMS. Working with the whole community to manage emergency incidents that may impact education agencies. Readiness and emergency management for schools REMS technical assistance center. Available from: https://rems.ed.gov/Resource_Plan_Basic_Community.aspx.

[11] United States Department of Health and Human Services Office of the Assistant Secretary for Preparedness and Response. 2017-2022 Health care preparedness and response capabilities, 2016. Available from: https://www.phe.gov/preparedness/planning/hpp/reports/documents/2017-2022-healthcare-pr-capablities.pdf.

[12] United States Department of Health and Human Services Office of the Assistant Secretary for Preparedness and Response. U.S. National Health Security Strategy 2019–2022, 2019. Available from: https://aspr.hhs.gov/ResponseOperations/legal/NHSS/Documents/NHSS-Strategy-508.pdf.

[13] United States Department of Health and Human Services Office of the Assistant Secretary for Preparedness and Response Office of Strategy, Policy, Planning, and Requirements. National Health Security Strategy Implementation Plan 2019-2022, 2019. Available from: https://www.phe.gov/Preparedness/planning/authority/nhss/Documents/2019-2022-nhss-ip-v508.pdf.

[14] Sobelson RK, Wigington CJ, Harp V, Bronson BB. A whole community approach to emergency management: strategies and best practices of seven community programs. J Emerg Manag 2015;13(4):349.

[15] Horney J, Nguyen M, Salvesen D, Tomasco O, Berke P. Engaging the public in planning for disaster recovery. Int J Disaster Risk Reduct 2016;17:33–7.

[16] Sherry N, Harkins AM. Leveling the emergency preparedness playing field. J Emerg Manag 2011;9(6):11–5.

[17] Kelman I, Stough L, editors. Disability and disaster: Explorations and exchanges. Springer; 2015 Jul 15.

[18] Americans with Disabilities Act of 1990, as amended, 42 U.S.C. §§ 12101 et seq.

[19] Americans with Disabilities Act Amendments Act of 2008, 42 USC 12101.

[20] Section 504 of the Rehabilitation Act of 1973, as amended, §504, 29 U.S.C. § 794.

[21] The Civil Rights Act of 1964, as amended, 42 U.S.C. § 2000d et seq.

[22] Anon. Patient Protection and Affordable Care Act of 2010, 42 USC 18116. Available from: https://www.law.cornell.edu/uscode/text/42/18116.

[23] U.S. Department of Health and Human Services Office of the Assistant Secretary for Preparedness and Response Office of Strategy, Policy, Planning, and Requirements. National health security strategy evaluation of progress, 2015–2018. Available from: https://aspr.hhs.gov/ResponseOperations/legal/NHSS/Documents/nhss-eop-508.pdf.

[24] U.S. Department of Health and Human Services Office of the Assistant Secretary for Preparedness and Response Office of Strategy, Policy, Planning, and Requirements. National Health Security Strategy Implementation Plan 2019–2022, 2019. Available from: https://www.phe.gov/Preparedness/planning/authority/nhss/Documents/2019-2022-nhss-ip-v508.pdf.

[25] Centers for Disease Control and Prevention. Building a learning community & body of knowledge: Implementing a whole community approach to emergency management, 2013. Available from: https://stacks.cdc.gov/view/cdc/31551.

[26] Gershon RR, Muska MA, Zhi Q, Kraus LE. Are local offices of emergency management prepared for people with disabilities? Results from the FEMA Region 9 Survey. J Emerg Manag 2021;19(1):7–20.

[27] Leser KA, Jetty A, Yates SC, Li J. NACCHO's baseline assessment of disability inclusion within local health departments. J Public Health Manag Pract 2016;22(5):496–7.

[28] National Association of City and County Health Officials (NACCHO). National assessment of the knowledge, awareness, and inclusion of people with disabilities in local health departments' public health practices, 2014. Available from: https://www.naccho.org/uploads/downloadable-resources/researchbrief_disabilityassessment_dec2014.pdf.

[29] Monterey County Office of Emergency Services. County of Monterey operational area emergency operations plan, 2020 Nov. Available from: https://www.co.monterey.ca.us/home/showpublisheddocument/97645/637393012001230000.

[30] Miami-Dade County Office of Emergency Management. Miami-Dade County, Florida comprehensive emergency management plan (CEMP), 2017. Available from: https://www.miamidade.gov/fire/library/OEM/CEMP.pdf.

[31] Individuals with disabilities and access and functional needs. Mayor Antonio Villaraigosa, Los Angeles Executive Directive No. 23 May 23., 2012. Available from: http://ens.lacity.org/mayor/villaraigosa/mayorvillaraigosa331283140_05232012.pdf.

[32] Cal EOS Governor's Office of Emergency Services. Assembly Bill 2311 planning guidance, 2017. Available from: https://www.caloes.ca.gov/AccessFunctionalNeedsSite/Documents/Guidance%20-%20AB%202311%20(6-15-17).pdf.

[33] Kailes JI, Enders A. Moving beyond "special needs" a function-based framework for emergency management and planning. J Disabil Policy Stud 2007;17(4):230–7.

[34] North Carolina Emergency Management. Access and functional needs toolkit for emergency managers, 2018. Available from: https://files.nc.gov/ncdps/AccessFunctionalNeedsToolKit_Final.pdf.

[35] FEMA. Glossary. U.S. Department of Homeland Security FEMA; 2021. [cited 2021 Dec 21] Available from: https://www.fema.gov/about/glossary.

[36] United States Department of Health and Human Services Office of the Assistant Secretary for Preparedness and Response. At-risk individuals. Public Health Emergency. U.S. Department of Health and Human Services Office of the Assistant Secretary for Preparedness and Response; 2021. [cited 2021 Dec 21]. Available from: https://www.phe.gov/Preparedness/planning/abc/Pages/atrisk.aspx.

[37] Association of University Centers on Disabilities (AUCD). Prepared4ALL whole community inclusive emergency planning [Internet]. Association of University Centers on Disabilities (AUCD); 2021 [cited 2021 Oct 21]. Available from: https://nationalcenterdph.org/our-focus-areas/emergency-preparedness/prepared4all/online-training/.

[38] Kailes JI. Defining functional needs – updating CMIST [Internet]. The partnership for inclusive disaster strategies; 2020 [cited 2022 Jan 5]. Available from: https://disasterstrategies.org/blog-post/defining-functional-needs-updating-cmist-by-june-isaacson-kailes-disability-policy-consultant/.

[39] Patel S. Engaging with and training community-based partners to improve the outcomes of at-risk populations after public health emergencies: findings from case reports [Internet]; 2019. Available from: https://www.nationalacademies.org/documents/embed/link/LF2255DA3DD1C41C0A42D3BEF0989ACAECE3053A6A9B/file/D374B8FEC9946913613EA440A8A882F0A59DCB4D56F5.

[40] Sammartinova J, Donatello I, Eisenman D, Glik D, Prelip M, Martel A, Stajura M. Local public health departments' satisfaction with community engagement for emergency preparedness. Am J Bioterror Biosecur Biodefense 2014;1:1–6.

[41] FEMA. Voluntary and community-based organizations [Internet]. U.S. Department of Homeland Security/FEMA; 2021 [cited 2022 Jan 3]. Available from: https://www.fema.gov/emergency-managers/individuals-communities/voluntary-organizations.

[42] FEMA. Organizations preparing for emergency needs (OPEN) online course [Internet]. Ready.gov. [cited 2022 Jan 2]. Available from: https://community.fema.gov/opentraining.

[43] U.S. Department of Health and Human Services Office of the Assistant Secretary for Preparedness and Response. Engaging community-based organizations [Internet]. Public Health Emergency; 2021 [cited 2022 Jan 3]. Available from: https://www.phe.gov/Preparedness/planning/abc/Pages/engaging-CBO.aspx.

[44] Koch H, Franco ZE, O'Sullivan T, DeFino MC, Ahmed S. Community views of the Federal Emergency Management Agency's "whole community" strategy in a complex US city: re-envisioning societal resilience. Technol Forecast Soc Chang 2017;121:31–8. https://doi.org/10.1016/j.techfore.2016.12.009.

[45] Novak J, Sorpory P. Engaging with and training community-based partners for public health emergencies: Qualitative research evidence synthesis. Washington, DC: The National Academies Press; 2020. Available from: https://www.nationalacademies.org/our-work/evidence-based-practices-for-public-health-emergency-preparedness-and-response-assessment-of-and-recommendations-for-the-field.

[46] Scott J, Coleman M. Reaching the unreached: building resilience through engagement with diverse communities. J Bus Contin Emer Plan 2016;9(4):359–74.

[47] Sievers JA. Embracing crowdsourcing: a strategy for state and local governments approaching "whole community" emergency planning. State Local Gov Rev 2015;47(1):57–67.

[48] Schoch-Spana M, Ravi S, Meyer D, Biesiadecki L, Mwaungulu Jr G. High-performing local health departments relate their experiences at community engagement in emergency preparedness. J Public Health Manag Pract 2018;24(4):360–9.

[49] National Association of County & City Health Officials (NACCHO). Follow-up national assessment of the practices, awareness, and inclusion of people with disabilities in local health departments' public health practices, 2018 Jun. Available from: https://eweb.naccho.org/eweb/DynamicPage.aspx?WebCode=ProdDetailAdd&ivd_prc_prd_key=e150c726-72ce-49a0-bd44-d5516b95a3d3&ivd_qty=1&Action=Add&site=naccho&ObjectKeyFrom=1A83491A-9853-4C87-86A4-F7D95601C2E2&DoNotSave=yes&ParentObject=CentralizedOrderEntry&ParentDataObject=Invoice%20Detail.

[50] Adams RM, Prelip ML, Glik DC, Donatello I, Eisenman DP. Facilitating partnerships with community-and faith-based organizations for disaster preparedness and response: results of a national survey of public health departments. Disaster Med Public Health Prep 2018;12(1):57–66.

[51] Quick KS, Bryson JM. Public participation. In: Christopher A, Torfing J, editors. Handbook on theories of governance. Cheltenham, England: Edward Elgar Publishing; 2016.

[52] International Association of Public Participation. Spectrum of participation. International Association of Public Participation; 2014. Available from: https://cdn.ymaws.com/www.iap2.org/resource/resmgr/pillars/Spectrum_8.5x11_Print.pdf.

[53] International Association of Public Participation. IAP2 resources. International Association of Public Participation; 2022. [cited 2022 Feb 28]. Available from: https://www.iap2.org/page/resources.

[54] Nabatchi T, Amsler LB. Direct public engagement in local government. Am Rev Public Adm 2014;44(4_suppl):63S–88S.

[55] Grisez Kweit M, Kweit RW. Participation, perception of participation, and citizen support. Am Politics Res 2007;35(3):407–25.

[56] Kweit MG, Kweit RW. Citizen participation and citizen evaluation in disaster recovery. Am Rev Public Adm 2004;34(4):354–73.

[57] Schoch-Spana M, Selck FW, Goldberg LA. A national survey on health department capacity for community engagement in emergency preparedness. J Public Health Manag Pract 2015;21(2):196–207.

[58] Cha BS, Lawrence RI, Bliss JC, Wells KB, Chandra A, Eisenman DP. The road to resilience: insights on training community coalitions in the Los Angeles County community disaster resilience project. Disaster Med Public Health Prep 2016;10(6):812–21.

[59] Wolf-Fordham S, Shea N. FEMA promising practices: closing gaps in local emergency plans and grassroots emergency planning [Internet]; 2015. Available from: https://adapresentations.org/webinar.php?id=21.

[60] Eunice Kennedy Shriver Center. UMass Chan Medical School, 2022. [cited 2022 Feb 1]. Available from: https://shriver.umassmed.edu/.

[61] Wolf-Fordham S, Shea N, Gleason P, Hamad CD. Revised by Wolf-Fordham S. Active planning workbook meeting the emergency & disaster needs of people with disabilities [Internet]. 2015, Revised 2021. Available from: https://nationalcenterdph.org/wp-content/uploads/2021/07/Emergency-Planning.pdf.

[62] Wolf-Fordham S, Shea N, Hamad CD. Active planning for mass care sheltering and evacuation of people with disabilities final project report; 2013.

[63] Jacobson PD, Wasserman J, Botoseneanu A, Silverstein A, Wu HW. The role of law in public health preparedness: opportunities and challenges. J Health Polit Policy Law 2012;37(2):297–328.

Building community resilience

9

Jill Morrow-Gorton

University of Massachusetts Chan Medical School, Worcester, MA, United States

Learning objectives

(1) Describe and compare the elements that make up a resilient community

(2) Explain the roles of people with disabilities and others with access and functional needs in community disaster management

(3) Explain the importance of and process for using tools to measure and improve community resilience and compare different community resilience measurement tools

Introduction

Disaster management increasingly focuses on individual and community resilience. Individual resilience promotes personal physical and mental wellbeing and assists a person in coping, adapting, and recovering in a disaster. As members of communities, resilient individuals contribute to the resilience of their community. Healthier, more resilient people create healthier, more resilient communities that are better able to manage a disaster. However, community resilience is not merely the sum or average of the resilience of its citizens. Rather communities comprise an amalgamation of complex social, economic, cultural and political systems that, while benefitting from individual community member resilience, can also harness the strength, power and resources of these systems in a way that makes the community as a whole greater than its parts [1]. Community resources are critical both during early disaster response before other resources arrive and in the aftermath during rebuilding. Community resilience fosters successful local implementation of the phases of disaster management: prevention, mitigation, response, recovery, and planning.

Defining resilience

In their *Special Report on Managing the Risks of Extreme Events and Disasters to Advance Climate Change Adaptation*, the United Nations' (UN) Intergovernmental Panel on Climate Change (IPCC) defines resilience as the "ability of a system and its component parts to anticipate, absorb, accommodate, or recover from the effects of a hazardous event in a timely and efficient manner" [2]. As the system in this definition, communities contain components in addition to their individual citizens. Businesses, faith-based organizations, providers of human and disability services, healthcare systems and local government make up some of the parts of a community that collectively share in its wellbeing. Whole community thinking seeks to engage these public, private, and non-profit sectors in a dialog to assess community capabilities and resilience in order to identify strengths and vulnerabilities [3]. Characteristics and vulnerabilities of communities help define their ability to be resilient and to respond to and recover from a disaster as well as ascertain capabilities that can be built to strengthen communities. Understanding what elements contribute to a resilient community can drive these activities and close gaps in capacity that create vulnerabilities.

Individuals and communities respond to events in varying ways as illustrated in the IPCC definition of resilience where the entity may absorb, adapt, or transform in response to an event or disaster [2,4]. Absorptive responses use resistance strategies to resist and block the stressor resulting in maintenance of the status quo. This response occurs on a daily basis as entities negotiate and absorb the impacts of every day events. Examples of this include fire, earthquake, or wind resistant buildings where a structure, built to withstand the effects of these events, sustains minimal to no damage from the event. In some events, however, entities may not be able to absorb the impacts, but will respond by adapting to it. With adaptation, change occurs through incremental adjustments in behavior allowing people to continue functioning without major changes in roles or outcomes. Examples of adaptation include changing farming techniques, rotating crops to decrease soil erosion or developing other sources of revenue as a result of a disaster. The differential impact of disasters on diversified strong economies such as that in New York City compared to weaker ones such as Haiti illustrates the effectiveness of these strategies to adapt to disasters [5,6]. In contrast to adaptation, transformation results in a substantive change in the nature, structure or function of an individual or community. These alterations may be technological, behavioral or cultural and examples include a household changing its source of livelihood or a region moving from relying on agriculture to exporting natural resources such as oil or hardwoods. Each of these responses can occur with the same disaster and households and communities may respond differently depending on their needs and abilities. The core features of resilience include the ability to learn, adapt, and transform when needed to maintain, recover or improve one's status.

Elements of community resilience

Aldrich and colleagues offer a concept of community resilience based on thinking about the nature of a community. They define community resilience as the collective ability of a geographically defined area such as a neighborhood or town to cooperate and efficiently manage a trauma and return to daily life [7]. In this definition, communities share both an identity and a sense of place. Silver and Grek-Martin describe that despite feelings of loss and grief over changes to familiar places in their community caused by a tornado, participants in their study expressed a strong sense of social cohesion resulting from shared experiences during and after the event [8]. The United Kingdom's Strategic National Framework on Community Resilience approached defining a disaster resilient community by describing the profile of a resilient community [9]. Briefly summarizing, the framework considers that resilient communities are made up of resilient people who not only use and adapt their existing skills and knowledge to prepare for and manage emergencies within their household, but also understand risks and partner with others including community champions to take their part in community planning, response and recovery for disasters. The detailed list of descriptors found in Table 1 provides a context for the role of the individual for themselves, in their community and with local and national emergency management in preparing and responding to a disaster.

Many different models and frameworks for conceptualizing community resilience exist, however, they tend to share a set of essential domains. These domains may be termed differently or thought about from different perspectives, but they represent similar concepts. The Community Capitals

Table 1 Characteristics of resilient communities.
1. People use and adapt their existing skills, knowledge and resources in planning, responding and recovering from emergencies or major events.
2. People are aware of their risks and how those link to local and national level risks. They understand how these risks may make them vulnerable and use that knowledge to actively prepare for the potential aftermath of a disaster.
3. People take steps to make their homes and families more resilient and recognize how to use their skills, experience and resources most effectively during an emergency.
4. People are actively involved in participating in decision making and discussion in order to preserve the assets and structures of their community.
5. Resilient communities have a champion who is a trusted member of the community and communicates with and engages and motivates others to be involved.
6. Resilient communities partner with local emergency management and other organizations before, during and after an emergency to help coordinate community resilience activities with that of emergency services.
Adapted from Cabinet Office Government of the United Kingdom. Strategic national framework on community resilience; 2011 March.

Table 2 Community capitals framework domains.
1. Human capital
2. Social capital
3. Cultural capital
4. Political capital
5. Financial and economic capital
6. Natural capital
7. Built capital
8. Preparedness capital
Adapted Emery M, Flora C. Spiraling-up: mapping community transformation with community capitals framework. Community Dev 2006;37(1):19–35.

Framework, developed by Emery and Flora, outlines seven core areas called "capitals" that contribute to community resilience [10]. Based on their review of the literature, Cox and Perry added another domain to these seven related to disaster planning and response. Many of the existing frameworks for disaster and community resilience are built on this capitals-based approach [11]. These eight domains, seen in Table 2, include human, cultural, social, political, financial/economic, natural, built and preparedness capitals. Each capital represents an area of community function that can contribute to community resilience and help a community develop into a thriving society. Taking a closer look at these will illustrate the importance of each capital to resilience, growth and functioning of the community.

Human capital

Emery and Flora conceived of human capital as the individual capacities of community members, including skills, knowledge, education, training, leadership and other resources [10]. Human capital is necessary for developing and mobilizing communities and for response in the event of a disaster. Studies show that the person most likely to provide immediate assistance in an emergency is a family member, friend, neighbor, coworker, or passerby rather than a professional first responder [12]. As well, these individuals may provide primary emotional and physical supports during the recovery process. In 2013 when two bombs exploded near the finish line at the Boston Marathon, spectators provided immediate physical assistance by creating make-shift bandages and tourniquets out of clothing and napkins from nearby restaurants and provided emotional support by staying with the injured victims until the professional first responders arrived. Enhancing the skills, knowledge and preparedness of individuals builds human capital and strengthens the ability to provide assistance, offering a valuable asset in a disaster or emergency [10]. Health and wellness, incorporating physical, psychological, social and spiritual wellbeing, constitutes another important dimension of human capital. All kinds of health, including physical health, are vital to resilience and healthier people form healthier communities. However, health limitations in one area may not always be associated with poor health in other areas. Populations of people with disabilities and other access and functional needs may have poorer physical health related to their disabling medical conditions, however, the need for accommodations to meet motor and other functional needs related to their disabilities does not necessarily reflect poorer mental, social or spiritual health [13,14]. People with disabilities often learn adaptability and flexibility through their life experiences giving them added psychological resilience which, if recognized, can be an asset to the community.

Human capital is also associated with other characteristics. A number of studies of children and adults that examined the qualities linked to their ability to adapt to adverse conditions demonstrated that certain factors increase one's ability to cope and adapt in a positive manner in the event of difficulty [12,15,16]. These qualities include personality factors, attitudes, coping skills, problem solving ability and social attachment. Hardy people with high levels of self-efficacy, self-esteem and self-reported resilience have been shown to have better post-disaster mental health outcomes. Attitudes of optimism and positivity, faith, and altruism also foster greater resilience and abilities to cope with disaster. Cognitive abilities and coping skills produce resilience by helping people use reasoning to solve problems and reduce stress. Lastly, the ability to connect with others socially to provide and receive social support as well as links to competent adults or peers supports resilience. Individuals with these strengths are more likely to engage in their community and in health promoting behaviors, while the literature shows that people with a poorer quality of life who lack these strengths engage less [17]. People with emotional and spiritual wellbeing and higher satisfaction with life also have greater participation in preparedness activities especially those related to evacuation. Thus, physical, psychological, spiritual and social health contribute to building human capital and resiliency in a community. Healthy people contribute to healthy communities by having more positive outcomes related to the presence of psychological strength, good physical health and quality of life, connections to family and friends and general well-being [15]. These all work together to enhance community involvement in the disaster cycle.

Social capital

Social capital builds on human capital and embodies relationships and how those impact cohesion and collaboration toward shared goals in a group [10]. The basic concept of social capital according to Lin is that individuals invest in and use resources in social networks to meet their needs [18]. Three fundamental social psychological dimensions of social capital and relationships, a sense of community, a sense of place, and citizen participation characterize resilient communities. Often shared interests and a common sense of place that lead to a social infrastructure comprised of networks and bonds are built into community relationships. Family cohesion, bonds with neighbors and friends, social networks and community engagement constitute some of the strengths of communities that add to their resilience. Longstaff recognized the importance of particular network members with many connections to others that can act as hubs linking between networks [19]. Social capital has been noted by many to be a key resilience element in disaster response and recovery [4]. Aldrich and Sawada studied the impact of resilience factors on deaths after the 2011 tsunami in Japan and determined that social capital had a greater role in lowering deaths from the tsunami than the height of the seawalls built to protect the population which had no apparent impact [20]. Successful adaptation and resilience in communities is reflected in community members' abilities to act as a group to coordinate decision making, take advantage of new opportunities and ways to innovate, and come together to actively engage and plan for the community as well as respond to disasters. Thus, the US Federal Emergency Management Agency (FEMA) instituted a whole community approach to disaster management in order to leverage the capabilities and strengths of communities, in particular their social networks and infrastructure, to help meet their needs in the event of a disaster [3,21].

Social connections, social groups, social networks, and social capital shape community resilience [21]. People belong to various groups within their family and community, including friendships, work and professional groups, religious affiliations, social groups, and neighborhood organizations [22]. These social groups vary in size from two to many people with structures varying from loose and casual to formal and structured with written rules. Social groups generally share a common identity which might be interests, values or belonging to the same social category. Researchers classify three types of social capital based on the strength of relationships and composition of networks: bonding, bridging, and linking. Bonding social capital occurs most often in people with a high level of similarities in particular interests or social level. Bonding social capital is associated with high levels of disaster response including the likelihood that individuals will receive warnings, prepare for disasters, evacuate, locate shelter and needed supplies, and obtain immediate and initial recovery assistance [23]. An example of bonding can be seen in the recovery of a flooded Vietnamese community in New Orleans after Katrina [7]. By sharing resources, coordinating activities and the support of the Catholic Church in the community, this tight-knit, highly bonded area, Village de L'Est, rebuilt more efficiently than less damaged, wealthier neighborhoods. Unlike bonding, bridging cuts across social groups such as race and class and results in loosely associated relationships with more diversity in people and resources. Bridging assists individuals in finding employment and other opportunities that would not be available to them within their own bonded group. It can result in the opportunities to share ideas and resources between groups, leading to innovation and tolerance. While bridging links are weaker than bonding ones, they provide resources and opportunities across social networks that are not available within them [24]. The third type of social capital, linking, can be viewed as a subset of bridging that occurs in a vertical rather than horizontal manner. Michael Woolcock advocates that this third type of social capital is needed to capture the sense of the power dynamic between social networks that influences a community [25]. Linking reflects the differences in power, social status and wealth between two groups where one group is clearly lower than the other. In contrast, bridging which can be between groups that have different power, wealth and social status, but they are equal in that one is not superior to the other. It functions to connect average community members with those in power and creates trusting relationships between networks of people across gradients of authority or power in society [7]. As a result, these three types of social capital create reciprocal connections between social networks and organizations with the ability to share resources and cooperate to make decisions which strengthen communities and build resilience [1].

One important aspect of community resilience is the ability of communities to link individuals to key social and other resources. Successful response to disasters relies on the ability of organizations and networks within a community to harness its social capital to meet the needs of the community and its members in periods of stability as well as during and after disasters. Townshend and others in their study of resilience in rural Canada highlight the importance of the connection between place-based social cohesion and resilience [26]. They note that social cohesion is made up of a number of characteristics including shared values and commitment (described as belonging) and inclusion (implying equal access, participation, recognition and legitimacy). These characteristics were evident during flooding and fire disasters in rural Canada where people helped their neighbors and community during both event response and community rebuilding. The US Centers for Disease Prevention and Control (CDC) defines social cohesion as the strength of relationships and sense of solidarity among community members [27]. One indicator of social cohesion is a community's level of social capital and shared group resources. Wulff and colleagues describe an example of social cohesion and resilience during

Hurricane Sandy that exemplifies how community resilience and social capital can support people with disabilities during a disaster [12,27,28]. A man with quadriplegia who used a ventilator to breathe lived in a home in one of the communities hit by Sandy. Over the 2 weeks of the hurricane and after, his friends and neighbors spontaneously partnered with the local fire station to keep his ventilator batteries charged while the electricity was out in the area. This effort allowed him to remain safely at home rather than having to move to a shelter, nursing facility or hospital to have his needs met. FEMA has capitalized on the concept of Neighbors Helping Neighbors to empower community leaders to educate their community about some simple steps to take to become more prepared for a disaster [29]. Their program includes a Community Preparedness Toolkit to help guide these activities.

Networking and partnerships in communities occur between community-based organizations (CBOs) that provide services in the community, individuals and groups. Also known as community- and faith-based organizations (CFBO), CBOs represent a number of different sectors including healthcare, social services such as disability services, cultural and faith-based organizations such as houses of worship and local arts groups, businesses, behavioral and mental health providers, education and child-care [30,31]. These CBOs not only interact with local, state and national government entities before, during and after a disaster, but also with each other. Similar to CBOs, the term, non-governmental organization (NGO), reflects an expanded view of organizations that can partner with each other or government to strengthen community resilience [32]. This concept adds both for-profit and nonprofit businesses to the other types of organizations considered as important partners in building community stability, responding to disasters, and strengthening the public health and medical systems. Local organizations such as CBOs and NGOs have strong networks with their communities through networks and grassroots activities and are invested in the long-term community wellbeing. They pull together community members and use community resources to collectively solve community problems. These entities have more freedom to meet community needs as they are not bound by the jurisdictions and rules that limit some governmental activities at the municipal, state or federal level. This makes them able to respond more flexibly to meet the need at hand as well as change approaches as the needs change. Typically more connected organizations deliver services more effectively than those with fewer strong ties. Thus, leveraging and strengthening social capital using existing networks, empowering local groups and networks to play an active part in their community, and enhancing relationships and connections between CBOs and other CBOs and government will contribute to building community resilience.

Cultural capital

Cultural capital, according to Emery and Flora, signifies a community's values, customs, languages and traditions [10]. The traditional sociologic approach to culture views it as including shared norms and values and a way of collective expression. This can include ways in which creativity is expressed and shared within the community. Coined in the 1960s by French sociologist Pierre Bourdieu, cultural capital took on a somewhat different meaning [33]. He noted that wealth or the lack thereof was not sufficient to explain educational disparities in learning and achievement between children of different social classes. Rather, he posited cultural capital as a sort of cultural behavior and endowment that denote power and status and contribute to social class as much as wealth [34]. Bourdieu notes that people are socialized by the people that they live with and around as well as by their environment [35]. This socialization molds people's tastes, preferences, choices and behavior, creating a lifestyle that is a marker of social position. Some of these characteristics include academic achievements and credentials such as degrees from prestigious schools, ways of speaking and manners including language and gestures, and cultural objects and possessions including fine arts and luxury brands. This dual meaning of culture conveys some of the features underlying social differences that give some groups more influence than others. Cultural capital differs from social capital in that it represents characteristics of individuals and social groups while social capital means the connections and networks between these groups. However, cultural capital can define the vertical nature of social networks and represents one of the reasons that some groups and networks are more powerful than others.

Political capital

Political capital refers to the ways in which communities are governed within a formal governance structure, such as a mayor and city council form of local government, and informally through organizations such as political interest groups [10,35]. For both formal and informal governance, this consists of governmental structures, systems and processes including citizen access to resources, communication, interactions between levels of government, and investments in services including emergency management and the elements that lead to population resilience. In the US, formal government is defined in the Constitution and laws of the country, operationalized through regulation, and dispute resolution, including determining the constitutionality of laws, by the court system. Westerwinter studied types of informal governance and how these interface with formal governance in multiple systems of governments across the world [36]. He describes informal governance as unwritten rules that may be imprecise and mutual expectations and norms that are not formally part of the government, but which can influence and change existing legal rules and regulations. Entities may use these informal methods as a way of showing power and influencing outcomes in their favor. This phenomenon occurs at all levels of government and involves varied issues. Researchers in Brazil studying circular economy, a process by which a product is consumed, recycled and consumed in a cyclical pattern, and green, sustainable practices in supply chain management found that both formal mechanisms such as contracts and environmental standards and informal ones using trust and cooperation between parties positively encouraged the adoption of green practices within the supply chain network [37,38]. In their conceptualization of political capital, Emery and Flora also viewed political capital as a function of the participatory nature of government including how community members participate in governance and decision making and how they are included, engaged, and able to have their voice and opinions heard and considered [10]. Community engagement is an important aspect of supporting emergency management and resilience.

Economic capital

Bourdieu focuses on three of the eight capitals: cultural, social and economic [39]. He envisions the cultural capital as educational qualifications, the social capital made up of social obligations and institutionalized as "titled nobility," like rulers, and the economic capital as land ownership and property rights. Property ownership represents one type of financial and economic capital that provides a stable home and business. Economic capital also refers to other types of material resources such as money, stocks, or other assets that are readily convertible into money [35,39]. Emery and Flora discuss economic capital as the source of a community's economic vitality or a state where people in a given area have good jobs, viable businesses, profitable investments and there are few people at the poverty level [10]. This includes resources for employment and making a livelihood, wealth, and economic diversity with the ability to use alternate resources to offset economic downturns and inequalities. The Organisation for Economic Co-operation and Development (OECD) defines community and individual economic well-being as a reflection of material living conditions and wealth including the ability to purchase and consume goods as well as to control resources [40]. Economic health with respect to the amount of economic capital an individual or community has is directly related to physical and mental health [39]. Those living at or below the poverty line with little in the way of economic resources tend to experience more stress and feelings of powerlessness, which in turn unfavorably affect their health, while those with more economic resources can adopt healthier lifestyles including healthy diets, leisure time for exercise, decreased levels of stress, and other behaviors that promote better health.

Natural capital

Natural capital can be thought of as a vehicle like financial stocks that brings benefit to people in the form of assets, goods and services [10]. Natural resources such as clean water and air, forests, rich soil, geologic assets in fossil fuels and living organisms providing a food supply represent some of the ecosystem services available from nature. These provide people with a means to have not only economic viability, but also live a healthy life. Water demonstrates this relationship as people need water for various activities creating resources for the community. Agriculture needs water and healthy soil in order to grow food products. Manufacturing processes often require water to be able to produce their goods. Running water can be harnessed to produce electricity to run machines and power lights and appliances in homes. Communities vary in which natural resources are available in close proximity to them as products such as oil are not found in every community. Rural communities tend to be more dependent on soil, water, fisheries and forests for their living and, therefore, at higher risk for economic difficulties when natural resources are depleted. As raw materials often need to be processed in order to be used, that activity can offer economic resources to multiple communities and not just those in proximity to the natural resource. For example, trees that are harvested in forested areas in Maine might be transported to a saw mill in Georgia to be cut into wood and then sent to North Carolina to be made into furniture. This natural resource thus provides economic benefits for multiple areas across the country. In addition to natural resources, natural capital constitutes the management of natural resources, status of the environment, weather and climate, and ecological systems. Management of natural resources for use in recreational activities also benefits the health of the individuals and communities near them by increasing physical activity. Impacts of poor natural resource management, however, can put nearby communities at risk. Soil and land degradation resulting from poor practices not only impacts the land use, but also puts it at higher risk for destruction in the event of a natural disaster. Hurricane Maria in Puerto Rico (2017) was such an event [41]. This extreme weather event exposed the population of people living near the coast to coastal flooding and other health hazards from the environmental degradation that had occurred over time. The storm left people without electricity for a significant period of time post disaster. In this case, the hurricane had a greater impact on the populations of people living in poverty and those with disabilities related to the lack of resources to support them during and after the event. Thus, natural resources can provide communities with benefits or risks, building resilience or creating vulnerability, depending on the management of those resources.

Built capital

A community's building infrastructure comprises its built capital [10]. This consists of not only physical buildings such as homes, schools, community buildings, businesses with manufacturing plants and office buildings, but also other infrastructure such as water and sewage systems, energy grids, transportation and road systems, and communication lines. The latter two serve to link within and between communities and are at particular risk during disasters. The built environment includes green areas and open spaces and its design may foster accessibility or be a barrier to it. The CDC notes that the built environment affects a person's physical activity habits and lack or inaccessibility of sidewalks, bicycle or walking paths contribute to a sedentary lifestyle [42]. This can result in poor health outcomes with conditions such as obesity, heart disease, diabetes, and cancer that can cause disability and shorten life. Universal design principles promote accessibility of spaces for people with disabilities as well as those without disabilities and contribute to improved health and social interaction. In addition to the importance of the everyday design accessibility for the built environment, structural design of buildings and communities can contribute to their ability to withstand disasters and minimize structural damage [43]. By collecting and using information about local risks and hazards, communities can consciously plan for and build an infrastructure to minimize damage and destruction from the types of events to which they are most susceptible. For example, communities with flood risk can be designed with elevated housing above the potential flood line to prevent the properties from floodwater damage [12]. That same design can create useable, accessible, walkable and wheelable community spaces that support the development of healthy physical activity habits and building and expanding social networks which enriches the community's social capital.

Impacts of disasters on housing are not equality distributed among social or economic classes. Earthquakes tend to destroy the bottom tier of housing in low income areas and the wealthy typically rebuild their housing first, as occurred after the San Francisco earthquake of 1906 [43]. The earthquake damaged about half the housing stock, to the point of being uninhabitable, and it took people living in low income areas years to find housing and caused many to move out of the area. Where the housing is rebuilt, especially with safety or other upgrades, the price of housing often rises to the extent that those with low incomes can no longer afford it. This phenomenon also occurred after the San Francisco earthquake and after the 1972 floods in Rapid City, South Dakota. Temporary housing is one solution to keep people in the community and prevent them from leaving the area after a disaster. However, temporary housing disrupts existing social networks as it is often not organized geographically in order to keep people within certain social networks together. As well, aid for temporary housing after a disaster does not promote rapid recovery in the same manner that aid for permanent housing does. Therefore, recovery should focus on rebuilding the community, including strengthening buildings against future events and consciously planning the design of the community to foster better health and increased social capital, while retaining an element of affordability.

Models show that rapid restoration of housing and jobs prevents migration out of an area after a disaster. The resilience of the built environment impacts jobs as well as homes after a disaster. Businesses depend on their employees being able to get to work, power to run their buildings and plants and telephone service to communicate within and between businesses. Other elements of the built environment that are often impacted by disasters and affect the rate of recovery include power and electrical outages, interruptions in telephone service and disruption of the roads and rails blocking traffic in and out of the area. For example, during the 1994 Northridge, California earthquake, the highways crumbled, apartment buildings collapsed, and power was out across vast parts of the city [43]. This earthquake destroyed or rendered unsafe thousands of buildings that had to be demolished. Studies of the earthquake showed that 58% of businesses were closed because of lack of power, 56% because workers were unable to get to work due to highway disruption, and 50% because there was no telephone service. This had an immense impact on the community. Building robust, redundant power systems that would be less likely to fail would strengthen community resilience for businesses, but also would benefit people who use battery powered durable medical equipment [12]. This type of system would have allowed the person with the ventilator to use the power in his building to keep his equipment charged. Including satellite internet connectivity in the design of the buildings represents another potential solution to prevent disruptions of internet service. This option, deployed during Hurricane Sandy, allowed people living in a high rise building to contact family and friends or needed services immediately after the storm.

Preparedness capital

The final domain, preparedness capital, added to Emery and Flora's work by Cox and Perry, establishes the investments made by a community in emergency management including disaster planning, recovery, mitigation and preparedness as one of the core domains of community resilience [11]. Preparedness incorporates all of the activities, processes, and capabilities developed as part of emergency management. Preparedness capital includes: (1) the plan development process and the plan itself; (2) capability for formal emergency response by emergency personnel such as police, fire fighters and emergency medical services; (3) hospitals and health system capacity; (4) local, state, regional and national emergency management resources; and (5) public health resources and community establishments such as CBOs, NGO, local businesses, human service organizations, and volunteers. This capital is designed to support the community in the event of a disaster and to harness each of the other capitals, as relevant. to help the community get through the disaster and recover as efficiently and effectively as possible.

Models illustrate community resilience as a function of economic factors, including financial stability, strong physical and structural built environment, and natural resources and social factors, such as human and social capital, cultural identification, and political capital. While all of these elements are necessary, social capital with the existence of robust social supports and connections forms the strongest element in determining community resilience [1]. Community member engagement in each of the disaster cycle's phases will help assure that the developed plan will consider the strengths and weaknesses of the community and identify how to fill gaps and build resilience in the community.

Disaster resilience and disability

Communities and emergency management often view people with disabilities and chronic physical and mental health conditions, as vulnerable populations especially with regard to disasters and disaster planning. Studies show that children and people with disabilities are two to four times more likely to be injured or die in a disaster than adults without disabilities [14,44]. People with disabilities experience social and demographic factors unrelated to their specific disability that lead to disproportionate impact by disasters [45]. These factors include high rates of poverty, low income and low rates of employment, poor living conditions including poor housing construction, and lack of access to services and information, which together result in adverse outcomes for this population. Many communities lack good jobs for people with disabilities that accommodate their disability but allow them to make a reasonable wage. People with disabilities who are employed often work at lower paying jobs and have more sporadic employment which limits discretionary funds that might be used for disaster preparedness [14]. Exclusion from participation in decision making was also cited as a contributing factor which reflects the historic stigma and deficits in the inclusive planning process in emergency management. Fjord and Manderson turn around the perception of the vulnerability of populations of people with disabilities and posit that creating an accessible built and social environment would not only promote the inclusion of people with disabilities in the community more effectively, but also meet the needs of all of the community [46]. They note that movement from a common good, individual approach to disaster response to an inclusive one would begin to address social and structural inequalities that disproportionally impact people with limited incomes, those who have chronic conditions, and use wheelchairs for mobility. Inclusivity and universal design have the potential to increase community resilience for the whole population, including people with disabilities.

Barriers to disaster response for people with disabilities both parallel and diverge from those of the general population. Studies of Hurricane Katrina found that people with disabilities and their caregivers were less likely to evacuate before the storm because they were physically unable to do so or were caring for someone who was unable to evacuate [13]. In another study of an Alabama community near a chemical weapons storage site, 9% of the households had a family member with a disability who would require assistance evacuating, but only 60% of those households felt that they had adequate help, including transportation, to evacuate. Other evidence that people with disabilities face barriers during disasters was demonstrated during an emergency drill where volunteers who were designated as people with disabilities in wheelchairs and with visual impairments were ignored and left behind or treated inappropriately by emergency responders [13]. These are all indicators that the experiences of people with disabilities are vastly different from those without. However, there are other areas in which people with disabilities face the same barriers as people without disabilities, but to a greater degree.

Post-Hurricane Katrina disruptions of housing, employment, medical care and transportation affected the entire population, but had a disproportionally negative impact on people with disabilities [13]. Fox and colleagues spoke with 56 Katrina survivors with disabilities and found that these disruptions plus an interrupted communication infrastructure severely impacted the ability of individuals with disabilities to live independently after the hurricane [13,47]. Not only was their recovery slower, but none of these participants felt that they had yet recovered from the event. This was mostly attributed to the lack of resources. People with mobility difficulties or chronic medical conditions were unable to get durable medical equipment to support their health needs and the paperwork to replace lost or damaged equipment was substantial. The unavailability of medical care and services, especially those related to mental and behavioral health, also impacted recovery. Added stressors included living in temporary, relatively inaccessible housing, limited options for medical care and other needs, lack of employment opportunities and negative coping strategies, combined to exacerbate existing health conditions. Much of existing accessible housing was destroyed. Since much of it had originally been standard housing modified to be more accessible, it required sizeable expense and effort to rebuild, and this also contributed to poor recovery. Loss of resources such as housing or equipment has been shown to correlate with psychological distress as seen in this population after Katrina. Furthermore, the accessible public transportation, critical for independent living for people with disabilities that are unable to drive, became nonfunctional. Back up transportation plans failed as they often depended on family, friends, or neighbors and many of the whom had evacuated to different geographic locations. These examples illustrate the barriers and difficulties that people with disabilities face as a consequence of not having their needs met post disaster.

Building community resilience for all including people with disabilities

Poor disaster recovery experienced by people with disabilities can partially be attributed to a lack of inclusive planning. People with disabilities may be exposed to more stressors and have worse health, economic, and social outcomes post disaster, but those effects can be mitigated and communities can become more inclusive and more resilient in the process. This effort requires more than inclusive disaster planning, however, that can be a start. Considering the eight capitals of community resilience and the interventions that have been shown to create resilience points the way to building communities that welcome and support all of their members. Building human capital, including better health, linking social networks and cultures, universal design and inclusive policies, building sustainable, accessible communities, and providing economic opportunities can lead to more resilient communities.

Ballan and Sormanti identified a number of ways to promote resilience in people with intellectual disabilities in the context of disaster management and planning [48]. These basic skills and activities, such as providing concrete explanations and using social stories, focus on building resilience by validating the disaster role of community members, including those with intellectual disabilities. Even though their work primarily concentrates on the population of people with intellectual disabilities, it is important to recognize that these principles apply to all people with or without disabilities. However, it is critical that people with disabilities, who make up about 15% of the population, not only are considered when planning for disasters, but have a seat at the table to help with this task. Thus, communities should actively engage members from all different social groups, including those with disabilities, in order to be able to best plan for disaster response and build resilience.

Disasters are stressful events for everyone and one way that Ballan and Sormanti [48] propose to build resilience is to recognize and acknowledge the contributions that community members make both during and after the event. In addition, relationships with people in the community who care about them are critical to promote resilience, which exemplifies the importance of social capital linking people and networks together. It is also important to understand what people need to foster adaptive coping with the event. Early mental health support and intervention has been shown to minimize psychological outcomes from disasters, including post-traumatic stress symptoms. Ballan and Sormanti propose four activities for people with intellectual disabilities to improve disaster planning and mitigation and build community resilience. These activities include a focus on developing problem solving skills, engagement in evaluating the factors and activities to help promote empowerment and autonomy for this population, participation in community boards to add their voice to the discussion, and involvement in disaster relief efforts. These activities mirror those elements in the community capitals resilience framework that are associated with more effective building of community resilience.

The activities recommended by Ballan and Sormanti also parallel and enhance the dimensions of social cohesion. These dimensions include: (1) belonging; (2) being included; (3) participating; (4) being respected and accepted for one's self even though different; and (5) legitimacy. These dimensions reflect the elements to build social capital, one of the strongest resilience measures [26]. Participation in the community has been shown to increase community cohesion, which in turn leads to neighbors helping each other during a disaster and in community rebuilding. This collective sense of belonging and the social cohesion that accompanies it are positively linked to resilience and geographically universal. As well, some of the traditional preparedness activities continue to help community members become prepared and contribute to individual and community resilience. Factors like having enough supplies to survive 72 hours and developing both an evacuation and family reunification plan lead to better prepared citizens [12]. However, building individual and community resilience in other ways can also foster better outcomes post disaster. Studies show that pre-disaster mental health issues and ineffective coping skills contribute to more psychological distress and poor mental health post disaster [49,50]. In contrast, psychological well-being was positively predicted by resilience and problem-solving coping skills which can be taught. Learning these skills could promote more favorable adaptation in the wake of a disaster, thereby promoting better mental health. Manderscheid's Wellness Model links the mind and body, identifying that a healthy outlook on life improves physical health [51]. This model differentiates between wellness and illness where wellness is the degree to which one feels positive about life and illness is the presence of disease. Both can coexist and wellness is enhanced by the ability to manage one's feelings and related behaviors, and effectively cope with stress. To the extent that one builds effective coping strategies, learns to manage stress, and participates in positive health behaviors such as healthy eating and physical activity, then one builds wellness and in building wellness builds resilience. These principles apply to people with disabilities and chronic physical and mental health conditions as well as those without. Wulff notes that healthy individuals contribute to the formation of healthy populations which lead to stronger communities with the ability to withstand and adapt positively to adversity [12]. Thus, building resilient communities through preparedness and everyday wellness has a positive effect on both recovery from disasters and` everyday life.

Measuring resilience

Measuring resilience enhances the understanding of the characteristics and factors underlying it and is important to identify the interactions between those factors and what is most effective in building resilience. It also provides information related to community resilience level and gaps indicating areas for improvement. Results help communities prioritize where to focus their energies and investment in order to maximize resilience. Communities differ in strengths and needs and numerical scores identify the magnitude of these as well as providing solid evidence of change. Measurement allows monitoring of progress toward improvement and assuring the maintenance of gains. Early in the study of resilience few tools existed to measure it. However, in the past decade, a plethora of tools to measure community resilience have emerged [11,52]. The designs of many of these tools fit the purposes of the developers and may not provide a broad or complete enough view of community resilience to be useful for performance measurement and improvement. Additionally, Clark-Ginsberg and colleagues note that much of the work in development of tools to measure resilience focuses on the validity of the questions and the tool but not the ease of use or relevance [52]. They observe that tools that don't meet the community's measurement needs or are complicated to administer or score are likely to be set aside rather than used to provide information for improvement. Thus, finding tools that meet community needs and are easy to administer is important to measure performance and prioritize improvements in resilience.

In its 2012 report, *Disaster Resilience: A National Imperative*, experts assembled by the National Research Council put together a picture of the current state of disaster resilience in the United States and direction toward building greater resilience [53,54]. Part of this pathway recognized the importance of communities in disaster resilience. In response to this, the Council convened a workgroup to develop a framework for resilience measures and indicators for use by communities. The workgroup noted that a tool would assist communities in assessing their baseline resilience in a number of areas as well as evaluating their strengths and needs and setting priorities and goals for allocating resources. This would also allow communities to use the baseline and subsequent measures in a quality improvement process such as the Deming Plan-Do-Study-Act (PDSA) cycle of improvement to develop and test interventions to improve resilience factors [55,56]. The National Research Council (NRC) workgroup identified four key components to measure related to resilience: vulnerable populations, social factors, critical and environmental infrastructure, and built infrastructure. These four key components mirror some of the capitals of community resilience framework posited by Emery and Flora and incorporate many of the factors of all eight of Emery and Flora's capitals within the NRC's four key components areas [10,54]. The key component, vulnerable populations, stands for the goal of meeting the needs of populations impacted by factors such as chronic health issues, health disparities, mobility and other disabilities, and socioeconomic status. Social factors are those that enhance or hinder community recovery and include social and cultural capital, education, political capital in governance, and financial structures and workforce (economic capital). Critical and environmental infrastructure embodies infrastructures including water and sewage, electricity, transportation, and communications as well as natural capital. Lastly, the built infrastructure represents the ability of the roads, bridges and buildings such as housing, businesses, schools, hospitals and emergency facilities, to withstand a disaster. Together these form the basis of a comprehensive look at the aspects of resilience that communities can measure to monitor their progress.

According to Cutter who provided the background information for the National Research Council workgroup, existing tools to measure resilience can be grouped into two types: top-down and bottom-up [54]. Top-down tools constitute those that that would be used by a government oversight body or academic contractor to provide an external review of a community. These tools generally require expertise to use and offer a picture of the community focused on evaluating performance and helping decision making. Examples of top-down tools include the PEOPLES Resilience Framework which measures local resilience, the Baseline Resilience Indicators for Communities (BRIC) which measures county level pre-existing community resilience for comparison with counties across the United States, and the ResilUS designed to assess recovery of critical community services after a disaster [43,54].

Bottom-up tools are surveys and self-assessments that provide communities with a tool to evaluate their own performance on various dimensions of resilience. One example of a bottom-up tool, the Toolkit for Health and Resilience in Vulnerable Environments (THRIVE) targets communities of color to help improve health outcomes and reduce health disparities. Another bottom-up tool, the Coastal Resilience Index, created by the National Oceanic and Atmospheric Administration (NOAA), uses a self-assessment scorecard that a community completes to determine potential functioning following a disaster or hazard. The hazard can be changed based on the specific risks of the community and the tool evaluates the community's emergency management plan including buildings, infrastructure and mitigation strategies. These types of tools help individual communities assess themselves, but make comparing communities difficult.

The PEOPLES framework illustrates the breadth and range of some of these tools and demonstrates a top down approach to measuring resilience. Developed by MCEER, an earthquake engineering research center, the framework contains seven elements for measuring disaster resilience [43,57,58]. The principle behind these elements is that they include both what protects and what needs to be protected. The name,

PEOPLES, is an acronym for these seven elements: Population and Demographics, Environmental/ Ecosystem, Organized Governmental Services, Physical Infrastructure, Lifestyle and Community Competence, Economic Development, and Social-Cultural Capital, including social networks and systems. These elements mirror those in other frameworks and provide guidance and focus for interventions for improvement. However, this framework measures performance in each of these dimensions across time and space. The model layers the results to create a representation of the community. That representation serves as the basis for determining the community's performance during and after any disaster event allowing for the determination of a resilience index that can include some or all of the seven elements in the framework. The framework considers the impact of direct and indirect damage, the time to recovery, the recovery trajectory and the extent of recovery, and measures harm at the local and regional levels, including the effects on global supply chains. The results can help community managers and stakeholders determine the most effective strategy to rebuild after a disaster and where to focus efforts and resources.

THRIVE represents a bottom up model of evaluating and building community resilience related to health disparities in communities of color [59]. Developed using existing research and input from a national expert panel, the THRIVE toolkit provides communities with a self-assessment tool to identify factors in their community that could be modified to improve health and reduce health disparities. The literature shows that communities of color experience the same kinds of health conditions as the general population such as high blood pressure and diabetes, however, these conditions are more severe and more prevalent in communities of color [60]. As well, these communities face a number of social, economic and community conditions that produce a negative health impact [61]. Poverty, lack of access to healthy foods, poor social supports, substandard housing and environmental exposures contribute to health disparities. However, use of the THRIVE tool has demonstrated that it can help communities identify community qualities and attributes that promote positive health and safety outcomes for the population as well as ways to improve those outcomes. For example, a few months into piloting the tool and based on the results of a self-assessment, a number of communities initiated farmer's markets and youth programs to provide access to healthy foods such as fresh fruits and vegetables for all and positive activities for youth, addressing two gap areas. THRIVE provides a structure for local decision makers to partner with community members, coalitions of businesses and human services agencies, and public health practitioners and planners to outline community strengths and weaknesses and pinpoint the community factors associated with poor health outcomes for people of color. These partners then engage in planning and implementing evidence-based activities to address factors that begin to close the health gap and remedy health disparities. The tool promotes change in how people understand and think about health and safety and moves it to a more evidence-based foundation. As well, it allows for building community capacity and resilience using the strengths of the community and builds social capital by encouraging links among community members and leaders. Wulff and colleagues maintain that health and community resilience are linked and advocate for an approach that considers health in all policies where activities that promote community resilience and protection from disaster (e.g., projects that reduce damage from flood waters and create open spaces in the community for physical activity and socialization) also promote community wellness [12].

Gilbert also endorses the importance of using a consistent base to measure resilience and that measuring resilience is critical for determining priorities for improvement and for identifying whether community resilience has gotten better or worse [43]. Another set of indicators often cited as a good example of evaluating particular aspects of resilience is the San Francisco Planning and Urban Research

Association (SPUR) tool developed to measure and track the ability of the San Francisco Bay area to recover from earthquakes. The SPUR process identifies specific goals for recovery and sets a timeframe in which those goals should be met. This process incorporates many of the principles of the S.M.A.R.T. criteria for project management attributed to Doran [62]. Drucker also used this model in his management by objective approach to guide the goal development by the criteria which the S.M.A.R.T. acronym represents [63]. Goals should be specific, measurable, attainable, relevant and time-bound in order to be successfully executed. One example of a goal might be to have all school repairs completed and students back in the classroom learning within 30 days of an event. Other specific areas could include restoring function to utilities, resuming patient care in hospitals or rendering other community buildings, such as the fire and police stations, habitable. In this approach, each of these activities would have a specific time goal to meet based on the extent of damage and required repairs. This kind of planning and simulation activity has been shown to result in financial savings for the community. For example, the Multihazard Mitigation Council found that spending $1 on preparation before an event such as an earthquake saved about $4 in damages and restoration after the disaster [43].

Clark-Ginsberg and colleagues observe that the development and testing for many community resilience measurement tools focuses more on the validity of the questions and the tool rather than their usability [52]. They describe that many tools are cumbersome to use risking not being used at all or results not being used to their best advantage as emergency personnel get mired in details. Consequently, they joined with the international humanitarian and development organization GOAL to develop a tool and toolkit that not only incorporates the elements of resilience, but also features ease of use. The Analysis of Resilience of Communities to Disasters (ARC-D) toolkit uses resilience science including the Sendai Framework to bridge between gathering data for research and providing information for the practical building of resilience in a community. The tool incorporates eight community resilience sectors: Education, Economic, Environment, Political/governance, Health, Infrastructure, Social and cultural, and Disaster risk management and aligns the measures with the four 2015 Sendai Framework Priorities for Action. These four priorities include understanding disaster risk, strengthening governance for managing disaster risk, disaster risk reduction as a way to build resilience and enhancing preparedness to optimize the response. The tool produces three scores for each individual area, each of the Sendai priorities, and a total score. The toolkit provides standard guidelines for rating each measure and accounts for the community's resilience characteristics and activities in each area including awareness, action, comprehensiveness and sustainability. The assessors rate each measure using a 1–5 rating scale where 1 represents minimum resilience and 5, full resilience. To give a sense of the gradations between the rating levels, the complete scale rating choices are illustrated in Table 3. By using this tool, the developers feel that communities will be able to measure their progress in each of the important dimensions and be able to understand what that measurement means and where to focus their efforts and resources to build a more resilient community.

Planning creates stronger, safer, and more resilient communities as well as saving lives and money. Helping communities understand their current levels of risk and potential impact from hazards assists them in taking responsibility for their hazard risks and enables them to identify strengths and weaknesses in their ability to respond to and recover from a disaster. Measuring levels of capacity and capability for the various resilience dimensions supports communities in their efforts to improve their preparedness but also the livability and vivacity of their community. It also allows for assessing the impact of different policies and approaches to identify the one that best fits the community

Table 3 Rating levels of community resilience.
1. Minimum resilience: little awareness of disaster resilience issues and no action to address them
2. Low resilience: some awareness, motivation and action, but action is fragmented and fails to see the bigger picture
3. Medium resilience: awareness and long-term actions exist, but are not linked to long term strategies and goals and do not address all of the aspects of the problem
4. Approaching resilience: actions are long term, linked to strategy and goals and address main aspects of the issue, but implementation is lacking
5. Resilience: actions are sustainable and long-term, linked to a strategy, address all aspects of the issue, and are rooted in and embraced by the community
Adapted from Clark-Ginsberg A, McCaul B, Bremaud I, Cáceres G, Mpanje D, Patel S, et al. Practitioner approaches to measuring community resilience: the analysis of the resilience of communities to disasters toolkit. Int J Disaster Risk Reduct 2020;50:101714.

and its hazard risks. Engaging the whole community in this endeavor promotes cooperation and the development of social networks, empowers people such as those with disabilities who might otherwise have been excluded to bring their experience and points of view to the table, and cultivates a culture of inclusion, thus strengthening community resilience. Communities vary in their geography, hazard risks, economic resources and other factors, however, Gilbert describes basic elements and principles that apply to resilience for all communities [43]. These include informed individuals and groups such as families and neighborhoods who know the risks to their community, understand how to reduce them, and are organized to use that knowledge to prepare for disasters. As well, communities need to have a strong infrastructure with robust essential health, education and other services, thoughtful land-use planning, and effective building codes and standards. Use of a tool to measure these and other aspects of resilience can help communities assess their needs, establish baselines and set priorities and goals, and guide them toward better allocation of resources and improved resilience to benefit all community members and ascertain success.

Conclusion

Models suggest that community resilience is a function of multiple factors representing different dimensions of a community. Important elements include strong social support, connections and networks as well as robust economic development, land and resource management, and infrastructure. The role of human and social capital in promoting the physical and psychological health of the population is critical as is the integration between government and community-based entities in shaping the community and its preparedness. Researchers and practitioners advocate for the participation from people across the full continuum of the community including minorities, people with access and functional needs such as those with disabilities, and other potentially vulnerable subgroups in order to build a disaster plan that will meet the needs of and build resilience within the community for all community members [12–14]. Using one of the existing frameworks to guide community and disaster planning assures that a community will address the essential areas needed to build resilience within its populations and physical structure. Measuring baseline levels of function within these areas provides the starting point with which to guide planning and resource allocation. Subsequent measurement can identify successes and areas in which more attention is needed, thus driving continuous quality improvement by learning from experience. Planning and mitigation activities not only save money and lives, but also can provide the opportunity to consciously redesign the physical structure within communities to create more inclusive, usable environments promoting healthier and greater resilience.

Questions for thought

(1) Choose a community, such as a town, city, county, or region, identify a hazard risk for that area and analyze its strengths and weaknesses based on the eight capitals of resilience.

(2) Think about the community you chose in Question 1. Propose interventions to address each of the eight capitals to improve community resilience.

(3) Outline a process to measure baseline resilience and the effectiveness of the interventions you chose in Question 2.

References

[1] Norris FH, Stevens SP, Pfefferbaum B, Wyche KF, Pfefferbaum RL. Community resilience as a metaphor, theory, set of capacities, and strategy for disaster readiness. Am J Community Psychol 2007;41(1–2):127–50. Available from: https://onlinelibrary.wiley.com/doi/10.1007/s10464-007-9156-6.

[2] IPCC. Managing the risks of extreme events and disasters to advance climate change adaptation—IPCC. Ipcc.ch. IPCC; 2019. p. 34. Available from: https://www.ipcc.ch/report/managing-the-risks-of-extreme-events-and-disasters-to-advance-climate-change-adaptation/.

[3] FEMA. A whole community approach to emergency management: Principles, themes, and pathways for action, 2011. Available from: https://www.fema.gov/sites/default/files/2020-07/whole_community_dec2011__2.pdf.

[4] Béné C, Newsham A, Davies M, Ulrichs M, Godfrey-Wood R. Review article: resilience, poverty and development. J Int Dev 2014;26(5):598–623.

[5] Heritage Land Bank. How to prevent soil erosion, 2017. Available from: https://heritagelandbank.com/announcements/news-events/how-prevent-soil-erosion.

[6] Baily MN. Can natural disasters help stimulate the economy? Brookings; 2011. Available from: https://www.brookings.edu/opinions/can-natural-disasters-help-stimulate-the-economy/.

[7] Aldrich DP, Meyer MA. Social capital and community resilience. Am Behav Sci 2014;59(2):254–69.

[8] Silver A, Grek-Martin J. "Now we understand what community really means": reconceptualizing the role of sense of place in the disaster recovery process. J Environ Psychol 2015;42:32–41.

[9] Strategic National Framework on Community Resilience. Cabinet Office Government of the United Kingdom; 2011.

[10] Emery M, Flora C. Spiraling-up: mapping community transformation with community capitals framework. Community Dev 2006;37(1):19–35.

[11] Cox R. 9. Measuring community disaster resilience: A review of current theories and practices with recommendations. International Safety Research; 2015 May.

[12] Wulff K, Donato D, Lurie N. What is health resilience and how can we build it? Annu Rev Public Health 2015;36(1):361–74.

[13] Stough LM, Sharp AN, Resch JA, Decker C, Wilker N. Barriers to the long-term recovery of individuals with disabilities following a disaster. Disasters 2015;40(3):387–410.

[14] Roth M. A resilient community is one that includes and protects everyone. Bull At Sci 2018;74(2):91–4.

[15] Abramson DM, Grattan LM, Mayer B, Colten CE, Arosemena FA, Bedimo-Rung A, et al. The resilience activation framework: a conceptual model of how access to social resources promotes adaptation and rapid recovery in post-disaster settings. J Behav Health Serv Res 2014;42(1):42–57.

[16] Bonanno GA, Diminich ED. Annual research review: positive adjustment to adversity—trajectories of minimal-impact resilience and emergent resilience. J Child Psychol Psychiatry 2012;54(4):378–401.

[17] Gowan ME, Kirk RC, Sloan JA. Building resiliency: a cross-sectional study examining relationships among health-related quality of life, well-being, and disaster preparedness. Health Qual Life Outcomes 2014;12(1):85.

[18] Lin N. Social capital. London: Routledge; 2011.

[19] Longstaff PH. Security, resilience, and communication in unpredictable environments such as terrorism, natural disasters, and complex technology. Harvard University; 2005.

[20] Aldrich DP, Sawada Y. The physical and social determinants of mortality in the 3.11 tsunami. Soc Sci Med 2015;124:66–75.

[21] Pfefferbaum B, Van Horn RL, Pfefferbaum RL. A conceptual framework to enhance community resilience using social capital. Clin Soc Work J 2015;45(2):102–10. Available from: https://link.springer.com/article/10.1007%2Fs10615-015-0556-z.

[22] Aldrich DP. Building resilience: Social capital in postdisaster recovery. University of Chicago Press; 2012.

[23] Hawkins RL, Maurer K. Bonding, bridging and linking: how social capital operated in New Orleans following Hurricane Katrina. Br J Soc Work 2009;40(6):1777–93.

[24] Claridge T. What is bridging social capital? Social Capital Research & Training; 2018. Available from: https://www.socialcapitalresearch.com/what-is-bridging-social-capital/.

[25] Woolcock M. Microenterprise and social capital. J Socio-Econ 2001;30(2):193–8.

[26] Townshend I, Awosoga O, Kulig J, Fan H. Social cohesion and resilience across communities that have experienced a disaster. Nat Hazards 2014;76(2):913–38.

[27] health.gov. Social cohesion—Healthy people 2030. Available from: https://health.gov/healthypeople/objectives-and-data/social-determinants-health/literature-summaries/social-cohesion.

[28] Clay PM, Colburn LL, Seara T. Social bonds and recovery: an analysis of Hurricane Sandy in the first year after landfall. Mar Policy 2016;74:334–40.

[29] Ready.gov. Neighbors, www.ready.gov. Available from: https://www.ready.gov/neighbors.

[30] Acosta JD, Burgette L, Chandra A, Eisenman DP, Gonzalez I, Varda D, et al. How community and public health partnerships contribute to disaster recovery and resilience. Disaster Med Public Health Prep 2018;12(5):635–43.

[31] Adams RM, Prelip ML, Glik DC, Donatello I, Eisenman DP. Facilitating partnerships with community- and faith-based organizations for disaster preparedness and response: results of a National Survey of Public Health Departments. Disaster Med Public Health Prep 2017;12(1):57–66.

[32] Acosta J, Chandra A. Harnessing a community for sustainable disaster response and recovery: an operational model for integrating nongovernmental organizations. Disaster Med Public Health Prep 2013;7(4):361–8.

[33] Weininger EB, Lareau A. Cultural capital. The blackwell encyclopedia of sociology; 2007 February 15.

[34] Coulangeon P. Cultural capital. The Blackwell Encyclopedia of Sociology; 2020 October 22.

[35] Pinxten W, Lievens J. The importance of economic, social and cultural capital in understanding health inequalities: using a Bourdieu-based approach in research on physical and mental health perceptions. Sociol Health Illn 2014;36(7):1095–110. Available from: https://onlinelibrary.wiley.com/doi/pdf/10.1111/1467-9566.12154.

[36] Westerwinter O. Formal and informal governance in the UN peacebuilding commission. In: Jakobi AP, Wolf KD, editors. The transnational governance of violence and crime. Governance and limited statehood. London: Palgrave Macmillan; 2013. https://doi.org/10.1057/9781137334428_4.

[37] ecpr.eu. The politics of informal governance, 2016. [cited 2022 Jan 9]. Available from: https://ecpr.eu/Events/Event/PaperDetails/27164.

[38] Cardoso de Oliveira MC, Machado MC, Chiappetta Jabbour CJ, de Sousa L, Jabbour AB. Paving the way for the circular economy and more sustainable supply chains. Manag Environ Qual 2019;30(5):1095–113.

[39] Bourdieu P. The forms of capital. In: Handbook of theory and research for the sociology of education. Westport, CT: Greenwood Press; 1986.

[40] Anon. OECD framework for statistics on the distribution of household income, consumption and wealth OECD; 2013.

[41] Morris ZA, Hayward RA, Otero Y. The political determinants of disaster risk: assessing the unfolding aftermath of hurricane Maria for people with disabilities in Puerto Rico. Environ Justice 2018;11(2):89–94.

[42] Anon. Healthy Community Design fact sheet series impact of the built environment on health what is the public health issue? 2011. Available from: https://www.cdc.gov/nceh/publications/factsheets/impactofthebuiltenvironmentonhealth.pdf.

[43] WBDG—Whole Building Design Guide. NIST Special Publication 1117: Disaster resilience: A guide to the literature, 2010. [cited 2022 Jan 9]. Available from: https://www.wbdg.org/ffc/nist/criteria/nist-spec-pub-1117-dis-res-guide-lit.

[44] UNDRR. Sendai framework for disaster risk reduction 2015-2030, 2015. Available from: https://www.undrr.org/publication/sendai-framework-disaster-risk-reduction-2015-2030.

[45] Stough LM, Kang D. The Sendai framework for disaster risk reduction and persons with disabilities. Int J Disaster Risk Sci 2015;6(2):140–9.

[46] Fjord L, Manderson L. Anthropological perspectives on disasters and disability: an introduction. Hum Organ 2009;68(1):64–72.

[47] Fox MH, White GW, Rooney C, Cahill A. The psychosocial impact of hurricane Katrina on persons with disabilities and independent living center staff living on the American Gulf Coast. Rehabil Psychol 2010;55(3):231–40.

[48] Ballan M, Sormanti M. Trauma, Grief and the Social Model: Practice guidelines for working with adults with intellectual disabilities in the wake of disasters. Rev Disabil Stud 2014;2(3):78–97. [cited 2022 Jan 9]; Available from: https://www.rdsjournal.org/index.php/journal/article/view/339.

[49] Abramson D, Stehling-Ariza T, Garfield R, Redlener I. Prevalence and predictors of mental health distress post-Katrina: findings from the Gulf Coast Child and Family Health Study. Disaster Med Public Health Prep 2008;2(2):77–86.

[50] Mayordomo T, Viguer P, Sales A, Satorres E, Meléndez JC. Resilience and coping as predictors of well-being in adults. J Psychol 2016;150(7):809–21.

[51] Manderscheid RW, Ryff CD, Freeman EJ, McKnight-Eily LR, Dhingra S, Strine TW. Evolving definitions of mental illness and wellness. Prev Chronic Dis 2010;7(1):A19. 20040234. PMC2811514. Epub 2009 Dec 15.

[52] Clark-Ginsberg A, McCaul B, Bremaud I, Cáceres G, Mpanje D, Patel S, et al. Practitioner approaches to measuring community resilience: the analysis of the resilience of communities to disasters toolkit. Int J Disaster Risk Reduct 2020;50, 101714.

[53] National Academies Press. Disaster resilience. Washington, DC: National Academies Press; 2012. Available from: https://www.nap.edu/catalog/13457/disaster-resilience-a-national-imperative.

[54] Nih.gov. Introduction. In: Committee on measures of community resilience: from lessons learned to lessons applied, resilient America roundtable, policy and global affairs, National Research Council. National Academies Press (US); 2015. Available from: https://www.ncbi.nlm.nih.gov/books/NBK285736/.

[55] Institute for Healthcare Improvement. Science of improvement: Testing changes. Institute for Healthcare Improvement; 2019. Available from: http://www.ihi.org/resources/Pages/HowtoImprove/ScienceofImprovementTestingChanges.aspx.

[56] deming.org. PDSA Cycle—The W. Edwards Deming Institute, 2022. [cited 2022 Jan 9]. Available from: https://deming.org/explore/pdsa/#:~:text=Also%20known%20as%20the%20Deming.

[57] buffalo.edu. Multidisciplinary Center for Earthquake Engineering Research (MCEER), Engineering resilience solutions: From earthquake engineering to extreme events, 2008. [cited 2022 Jan 9]. Available from: https://www.buffalo.edu/mceer/catalog.host.html/content/shared/www/mceer/publications/MCEER-08-SP09.detail.html.

[58] Cimellaro GP, Renschler C, Reinhorn AM, Arendt L. Peoples: A framework for evaluating resilience. J Struct Eng 2016;142(10):04016063.

[59] Davis R, Cook D, Cohen L. A community resilience approach to reducing ethnic and racial disparities in health. Am J Public Health 2005;95(12):2168–73.

[60] CDC (2005) Health disparities experienced by black or African Americans - - - United States [Internet]. Available from: https://www.cdc.gov/mmwr/preview/mmwrhtml/mm5401a1.htm.

[61] Smedley BD, Leonard Syme S, Institute Of Medicine (U.S.), Committee On Capitalizing On Social Science And Behavioral Research To Improve The Public's Health. Promoting health: intervention strategies from social and behavioral research. Washington, DC: National Academy Press; 2000.

[62] Doran GT. There's a S.M.A.R.T. way to write management's goals and objectives. Manag Rev 1981;70(11):35–6.

[63] TechRepublic. Use S.M.A.R.T. goals to launch management by objectives plan, 2005. [cited 2022 Jan 9]. Available from: http://www.techrepublic.com/article/use-smart-goals-to-launch-management-by-objectives-plan.

Promising practices in disability-inclusive disaster management

10

Jill Morrow-Gorton[a] **and Susan Wolf-Fordham**[b,c]

[a]*University of Massachusetts Chan Medical School, Worcester, MA, United States,* [b]*Consultant, Association of University Centers on Disabilities, Silver Spring, MD, United States,* [c]*Adjunct Faculty of Public Health, Massachusetts College of Pharmacy and Health Sciences, Boston, MA, United States*

Learning objectives

(1) Explain the differences between promising practices, best practices and evidence-based practices.
(2) Explain strategies to promote inclusive disaster management and planning.
(3) Compare important elements of projects designed to promote inclusive disaster management and planning.

Introduction

Increasing equity and access throughout the emergency management cycle, and evolution toward whole community emergency planning and response in practice, requires systems change. The US emergency management system is a "system of systems," each with numerous parts. Effecting change in any one part of those systems through the creation, modification or shift in a program, policy or practice may lead to change in other parts of the system. Over time, sustainable change becomes a promising practice in the field. "Promising practices" represent those undertakings that show potential, i.e., measurable positive results, but have not been fully tested and do not have enough evidence of efficacy across a wider span of uses. "Best practices" on the other hand are those system changes with measurable results that become generally accepted over time and are adaptable to different situations [1]. Documentation of these practices may be found in the academic or gray literature which represents government documents and publications by non-profit organizations, foundations, and others. In contrast, evidence-based practices have undergone rigorous scientific research and show reproducible results documented in the scientific literature. The nature of the field of emergency management and the variations in types of disasters and communities across the globe make rigorous, controlled study difficult, so most successful, more widely applicable approaches constitute best practices. Promising practices may become best practices over time through ongoing implementation, evaluation, refinement and successful replication. As the disability and disaster field is somewhat new, the academic and gray literature outlining best practices in the field is limited. However, there are a number of promising practices which may become best practices over time.

The promising practices to be illustrated here represent those that promote inclusion and better outcomes for people with disabilities including those with disability related to mental health conditions as well as chronic medical conditions and developmental disabilities. These represent practices and activities from across the world focused on inclusion and improvement in outcomes. In order to consider them in an organized fashion, they have been grouped accordingly: structural, regulatory and policy changes and inclusion of people with disabilities and others with access and functional needs including self-advocacy and empowerment, behavioral health, and resilience building. Some practices focus on individuals while others address the needs of communities, nations or regions comprised of multiple nations. Many practices involve engagement with the emergency management system at multiple levels, such as local communities or at the national or regional level. Others represent partnerships between academia or community-based organizations with people with disabilities and others with access and functional needs and emergency management systems. The guidance provided by the United Nations (UN) regarding disability-inclusive planning and disaster risk reduction that is informed by current standard practices as well as promising practices supports reaching the goals of involvement of all citizens in this endeavor.

Structural, regulatory and policy changes

Structural change refers to the re-arrangement of and relationship between the parts or elements of a system. In emergency management this often involves changes in roles and responsibilities within the emergency management system and shifts in the way the system works based on those changes. Structural changes also include changes to the common framework used to conceptualize and operationalize managing emergencies and in this case changes to traditional emergency management and public health preparedness. System changes may originate from a number of sources that occur within a government or governing body of an entity. Most governments and other entities with governing bodies, including businesses as well as United Nations (UN), agencies employ a set of principles under which they strive to function. These principles drive their work and often are written down or captured as policy. Policy helps guide decisions and can be used to create a system of rules that govern the entity. In the government of nations, states or provinces, these rules are encapsulated in laws or legislation. New or amended federal, state and local legislation can change roles and responsibilities in the national, state or local emergency or public health preparedness system structure. Regulations constitute a tool for administrative entities within government to interpret and operationalize a law. Through regulations the administrative entity creates the rule about how a particular law will be implemented. Theoretically, policy, law and regulation all align allowing the guiding principles to become the rules of how activities such as those in emergency management or public health occur.

Changes in government and the administration of emergency management and public health functions do not solely result from laws and regulations although they are typically driven by policy. Disability-inclusive policies and priorities can lead to meaningful changes that do not require formal legislation. For example, changes to emergency plan content to incorporate inclusive planning and response issues or strategies to increase stakeholder engagement in local and state emergency and public health preparedness planning can be accomplished without legislation. Although for the latter, legislation may set the structure for the composition of advisory groups or government agencies. A formal statute may provide strong impetus for making needed structural changes. While government administrations take different approaches to crafting the rules for the system, some prescriptive and others philosophical, the desired end result is to put the guiding principles in action. To that end, policy, legislation and regulation can drive structural changes to promote disability-inclusive practices in disaster management.

Globally, the Third UN World Conference held in Sendai, Japan in March 2015 set the guiding principles into place for disaster risk reduction for 2015 to 2030 via the Sendai Framework [2]. Based on this action and framework, the United Nations Office for Disaster Risk Reduction (UNDRR) and World Health Organization (WHO) help nations implement the framework's principles in their country or region. In parallel with this instrument, the UN has other initiatives to address disability issues that many of the world's nations have committed support [3]. The Convention on the Rights of Persons with Disabilities (CRPD) seeks to change attitudes and approaches to people with disabilities from "objects" that need to be cared for to people with rights and responsibilities who are part of their communities. This convention outlines the principles and areas where people with disabilities have traditionally been excluded as well as adaptations that may be needed to assure full accessibility and inclusion. Disaster management and risk reduction are identified areas where people with disabilities have not been part of the active planning and response and the UNDRR and WHO are tasked with changing this situation. In response to the Sendai Framework and the work of the CRPD, UNDRR, WHO and others, including community based and non-governmental organizations (CBO, NGO) and many nations, have adopted new legislation to help drive these principles of disability inclusiveness.

The World Bank offers a review of some promising practices in disability inclusion for disaster management from across the globe [4]. The government of Indonesia passed a law, Disaster Management Law 24/2007, to broaden the response to disaster risk reduction. In order to spread this concept to the local governments, the government sought to develop Technical Assistance and Training Teams

(TATTs) to strengthen the knowledge and understanding of disaster risk reduction at the local levels [5]. Indonesia partnered with Arbeiter-Samariter-Bund (ASB), a German organization focused on civil protection, rescue services with rapid response teams for large scale emergencies and social services for children and adults, to develop a training program for the Local Disaster Management Offices (LDMOs). The program included teaching inclusive planning practices at the national, provincial and local levels [6]. The training promotes disability inclusive practices and provides guidance and opportunities for sharing experiences and best practices in order to help local personnel better shape inclusive disaster risk reduction. It also offers opportunities for working on planning with stakeholders. One outcome of this initiative was the development of a Disability Inclusion Service Unit for Disaster Management within a Central Java LDMO. This unit serves to help districts build inclusive programs, replicate and scale practices with positive outcomes as well as to provide data for analysis and monitoring and evaluation of the programs. Thus, Indonesia employed a number of structural, policy and legislative levers in order to promote disability inclusive disaster risk reduction throughout the levels of national, provincial and local government.

In the United States (US), the Americans with Disabilities Act of 1990 (ADA) provides protections against discrimination against people with disabilities and guides the provisions for accessibility of public places as well as the expectation for reasonable accommodations in the workplace to allow people with disabilities to work [7]. This law provides the underpinning for disability related practices and formally recognizes the rights of people with disabilities to be active members of their communities. It also requires that their states and communities provide the accommodations to enable them to fully participate. Using this ADA legal action and others as a backdrop and on-going activities and focus on the inclusion of people with disabilities in disaster planning from both public health and emergency management standpoints, Los Angeles County, California Department of Public Health (LACDPH) identified variations and gaps in disability-inclusive planning for emergencies [8]. To address this issue, LACDPH brought together a consortium of stakeholders and experts to identify how it could more effectively engage and include people with disabilities in the planning process in order to develop consistent and comprehensive plans to meet their specific needs. One goal of the project was to increase access to resources and tools that educate and guide practitioners in understanding the importance of and how to include people with disabilities in planning. The interventions included job aids or job action sheets designed for specific positions within the Incident Command System (ICS) including positions in operations, planning and policy. In addition, the project assessed and mapped disability-inclusive needs and activities across the emergency management and public health systems to better understand and address those needs.

In general, the ICS system doesn't have a clear place for responsibility for populations with disabilities and others with access and functional needs, which makes it easy for responsibility for the needs of this population to fall through the cracks. However, the LACDPH project addressed the need for identifying someone with responsibility for this population within its ICS system by developing a functional needs unit leader position within its disaster sheltering branch as part of operations. Adding this position to its ICS structure provides a focus for maintaining equal access in disaster shelters, including ensuring that people with disabilities are sheltered in mass care (general population) shelters and with needed adaptations to maintain their independence while staying there. As one of the participants in the LACDPH project, New York City Health Department (NYC Health) highlighted an example of a structural change they used close a gap related to identification of, outreach to, and community engagement with populations with access and functional needs. NYC Health expanded the responsibility of its ICS Public Information Officer (PIO) function to include identifying, creating messaging for, and outreach to populations with disabilities and others with access and functional needs. These expanded responsibilities now include outreach to disability and other local service provider agencies, creation of a community engagement team, development of messages targeting specific communities with access and functional needs and in languages other than English, in addition to the usual PIO functions of being the communication conduit within the ICS system and between the system and external audiences. Both of these examples of structural change within the emergency management and public health systems enhances the ability of these systems to reach, engage, and include populations of people with disabilities in planning and other disaster management functions.

Written emergency plans are one of the tools that emergency management uses in addressing disasters and the maintenance and updating of these plans is generally the responsibility of the government emergency management entity. While plan content may not be regulated, generally a format or framework is used to guide what goes into the plan. This format and content can be structured in a manner to promote the inclusion of disability specific information in the plan. Revising emergency and public health preparedness plans to reflect an inclusive philosophy and specify equitable practices for implementation during emergency planning, response and recovery promotes more inclusive planning. One way to steer plans toward consistently including this information and to determine the extent to which they do is to compare the plan to a checklist. Two examples of checklists for key elements needed to meet the needs of those with disabilities and access and functional needs include one created by Disability Policy Consultant June Isaacson Kailes and another list jointly developed by the International Association of Emergency Managers (IAEM) and the National Emergency Management Association (NEMA) [9,10]. These types of checklist have been used to study the completeness of plans regarding certain key elements, however, this method can be used to measure any set of elements in plans including promising practices. Fox and colleagues in their study titled *Nobody Left Behind* identified that only about 20% of a sample of 30 emergency management plans contained different ways to evacuate people with physical disabilities and that 67% of the emergency managers responsible for those plans had no intention of adding those elements [11]. In a follow-up review of 12 of those plans, none of them had those elements in the plan or policies, guidelines, or practices that were specifically designed to assist people with mobility impairments. Another way that checklists have been used is for the identification of promising practices. In an unpublished study, Wolf-Fordham used a checklist compiled from the Kailes and IAEM/NEMA checklists, legal decisions, scholarly and gray literatures to review 70 local emergency or public health preparedness plans or guides that purported to be inclusive and identify promising practices in disability-inclusion in the plans [12]. Promising practices were then compiled from the individual plans, checklists, and literature into a list of 18 unique promising practices. This strategy outlines a way to extract promising practices and elements from plans in order to be able to share and learn from them.

Another tool with a number of uses that can be considered a promising practice is geomapping. This technique takes raw data from surveys or other methods of information gathering and puts that information on a map of an area. In public health, this type of mapping melds medical information with geography and informatics to analyze infectious disease outbreak locations or trends in conditions such as cancer [13]. In the United States, the federal Department of Health and Human Services (HHS) created an interactive map that identifies the locations of areas with concentrations of technology dependent Medicare beneficiaries who are generally elderly or disabled [14]. This geomapping application called the "HHS emPOWER Map" shows the total number of technology dependent Medicare beneficiaries who rely on electricity to power and charge their durable medical and assistive equipment and devices that allow them to live at home independently rather than in institutions. Only aggregated anonymous information is available to the public to protect individual confidentiality; during disasters under certain circumstances a public health authority may access more specific information in the database to use in emergency response. This information can inform disaster management especially in those areas where the risk for power outages is great and there are substantial numbers of people with electricity-dependent devices. Others have applied this strategy as well. Ecuador, in partnership with some of the surrounding countries, developed a process to geo-locate citizens with disabilities in areas where earthquakes are common in order to be able to track and assure that they have been safely evacuated to an accessible shelter in the event of an earthquake in their area [4,15]. This method to identify the geographic locations of people with disabilities and access and functional needs who may require accommodations in order to safely evacuate, support for power outages or other occurrences that would have a significant impact on their health and well-being can be applied without the technology as well. In its initiative to promote inclusive planning, Bangladesh created a project called the Community Based Disability Inclusive Disaster Risk Reduction project in the Gaibandha district which is an area prone to annual flooding from the Brahmaputra River [4]. In this project emergency planners worked with local groups that included people with disabilities from the affected communities to identify ways to reduce the risk from flooding. As part of this activity, a map of the area was drawn and people identified where the elderly and people with disabilities lived as well as the areas that were safe for evacuation. This mapping employed the same concepts as the emPOWER map, but based on the local knowledge of community members.

As with the emPOWER map, the Gaibandha district community members geographically located people with disabilities in the district in order to plan to meet those needs when a disaster occurs. This concept represents a practice that can be used in a variety of settings with varying resources which moves it closer to being a best practice.

Inclusion of people with disabilities and access and functional needs

About 15% of the world's population is estimated to have a disability and as people age the proportion of those with a disability or access and functional need increases [4]. Disasters take a greater toll on these populations as their mortality rates can be many times that of the general population. They experience greater morbidity related to exacerbation of existing health conditions and the development of new ones. As well, they undergo more extreme socio-economic disadvantages through disruption of services and a longer road toward disaster recovery. The Sendai Framework for Disaster Risk Reduction 2015–2030 and work from many nations across the world sets the stage for making inclusive practices mainstream [16]. In addition, evidence points to the value of disability-inclusive processes not only to meet the needs of those with identified disabilities, but to support those in the general population that may need occasional assistance such as pregnant women or who those who may have diminishing skills such as older adults [17]. Disability-inclusive design or universal design functions to create entrances that are easier to navigate for children and elderly as well as people with disabilities. Newer buildings and evacuation shelters in India and Bangladesh illustrate this idea. In Romania, citizens advocated for making fire stations accessible because people with disabilities might need to access these public buildings and building use changes over time. They reasoned that while currently the fire fighters in the building might not have disabilities people working in the building in the future might. What today is a fire station, tomorrow might be used as a community center and making it accessible from the start makes the building multifunctional. As well, universal design has been shown to only add a marginal increase to cost when building accessible structures from the beginning as compared to reconfiguring buildings to add accommodations. One study from Australia demonstrated that retrofitting buildings added 6% to the cost while building accessible from the start only added 0.2% [18]. However, in order to accomplish such changes in thinking and approaches, people with disabilities must be at the planning table from the beginning.

Stakeholder participation in political or non-political activities aimed at improving the community or addressing issues of public concern is a form of civic engagement. Civic engagement is a well-known and widely accepted practice. Advocates and scholars concur that public participation in emergency management and public health preparedness is critical to advance equity in the field [19]. A number of efforts have involved increasing stakeholder engagement in the emergency planning process. The well-known disability slogan, "Nothing About Us Without Us," sums up the importance of the people impacted by a policy or decision having a say in that decision or policy [20]. The question is how to get people to the table not just once or twice, but on a regular basis as established participants in the planning process. One practice is the establishment of functional and access need or disability advisory committees to advise the planning and emergency management process.

Various states in the United States developed projects that hinge on the use of functional and access need or disability advisory committees. In Washington State, the Washington State Coalition on Inclusive Emergency Planning (CIEP) is tasked with promoting independent living for people with disabilities in the state [21]. The CIEP began in 2015 and is funded by the state and administered by the Washington State Independent Living Council (WASILC), a governor-appointed board, developed, managed by, and serving people with disabilities. The WASILC partners with the Centers for Independent Living (CIL) or Independent Living Centers (ILC) as they are known in other states. These are entities formed under the federal Rehabilitation Act of 1973 [22]. CILs/ILCs are agencies controlled and run by people with disabilities and tasked with providing information and referral, supporting transition from institutions to the community, advocacy, peer counseling and training for people with a broad range of disabilities [23]. The CIEP describes itself as a cross-disability network drawing from independent living principles to provide technical assistance to assure that equity and disability inclusion become part of all facets of emergency management. In addition to providing expertise in accessibility and meeting the needs of populations of people with disabilities, CIEP plays an active role in disasters. It activates during a disaster upon notification by the state, the American Red Cross, other state or local agencies and/or stakeholders and there is a formal protocol for CIEP to communicate with its stakeholders and partners to share or receive information, highlight issues, and seek resolutions to problems that arise. In addition to participating in emergency response, CIEP also monitors recovery efforts in order to provide input into post-disaster reviews known as After Action Reports (AAR). During the COVID-19 pandemic, CIEP provided technical support related to COVID educational videos in American Sign Language (ASL), performed Americans with Disabilities Act (ADA) assessments of quarantine facilities, public testing and vaccination sites, and addressed equity-related issues [24]. The building of disability coalitions to engage with emergency management systems represents a promising strategy for integrating disability-inclusion into emergency management.

In another promising practice, the Council of Persons with Disabilities Thailand participated in an emergency management exercise with the Thailand-Cambodia Joint and Combined Exercise on Humanitarian Assistance and Disaster Relief [4]. The goal of the exercise was to improve and better coordinate disaster response between Thailand and Cambodia and within Thailand. Members of the

Council with disabilities filled the roles of a liaison officer in the Commanding Post Unit and others played the role of civilian disaster "victims" in a number of different drill events such as those simulating tsunamis and storm surges, landslides and collapsed buildings. By participating in these exercises, the other participants interacted with the members of the Council, learned about how people with disabilities experienced various disasters and were motivated to restructure their emergency management practices to be more inclusive.

In the 2000s, Ethiopia experienced difficulties with food security related to droughts. In order to address more rapid recovery from droughts, an Ethiopian group, the Gayo Pastoral Development Initiative, in the Borana district worked with the Intermon Oxfam to improve access to water and better food sources [4]. By building new or restoring existing water pools to increase access to drinking water, providing seeds for food that need less water to grow, and better quality animals for food sources, the initiative would reduce the risk of starvation in the event of another drought. The group engaged and paid people with disabilities to participate in activities including digging pools and restocking livestock. Those that couldn't physically take part in the activities did other tasks or had family members that filled their spots. This project not only allowed people with disabilities to have an active role in building their community and its resilience to drought, but also gave them an equal place in and opportunity to contribute to the community. This approach helps include people with disabilities in the community and empowers them to participate in a meaningful manner.

Part of effective participation as a stakeholder requires understanding the role, the people that are involved, and being able to communicate thoughts and ideas about the topic. The development of tools and toolkits to support inclusive planning and also to help disability coalitions and representatives to be effective members of planning teams represents another common practice with positive results. One example is the Association of University Centers on Disabilities' (AUCD) Prepared4ALL initiative, a project to build capacity for local disability organizations, emergency management, and public health preparedness agencies to make outreach, engage and collaborate with each other [25]. The project contains materials and tools for disability organizations and their networks to use to build those connections with emergency and public health professionals with the goal of building trust, creating working relationships, and getting a seat at the emergency planning table. Prepared4ALL focuses on capacity building among disability organizations, local emergency management, public health agencies, and other partners to collaborate regarding local planning and response. The goal of Prepared4ALL is building capacity for engagement and collaboration, ultimately leading to greater equity and inclusion in local emergency planning. Project research, including an online survey of emergency managers and public health preparedness planners, indicated that people with disabilities and disability organizations may lack knowledge and context about how the American emergency management system works, while emergency managers and public health preparedness planners may lack knowledge and context about the "why" and the "how" of whole community emergency management [26].

The Prepared4ALL initiative includes an online course which addresses the knowledge and experience gaps identified in the disability community and among emergency management and public health. In addition to the engagement and collaboration process, Prepared4ALL focuses on topical content based on the idea that relationship building is critical to the whole community process, an important point made by advocates and practitioners [27,28]. To that end the project team developed nine outreach, engagement and collaboration strategies, accompanying online course modules and related tools and resources, plus a strategy playbook. The strategies are based on several well-known behavior change approaches, including appreciative inquiry, action research, and adaptive thinking [28].

Stakeholders are key decision makers in these positive strengths-based approaches which involve trial and error problem solving and building on what works. The pinwheel in Fig. 1 illustrates the nine strategies developed by the project team [29].

- P is for Pinpoint. Pinpoint the disability inclusion issue at hand and identify the relevant local emergency management and public health agencies.
- R is for Related: Find local organizations with related goals for collaboration.
- E is for Engage: Along with your partner organizations engage the local emergency management and public health agencies. Show them your interest in their work and what your organization can offer to them.
- P is for be Positive: Focus on the strengths of your organization, your partners, and the local community in solving problems. Focus on what you can do and accomplish, not what you can't. Identify ways to fill in gaps.
- A is for take Advantage: Take advantage of opportunities and timing as you move forward.
- R is for Reflect as you plan. Don't get stuck by "we have no time," "we have no money," "we don't have enough staff" responses from local emergency managers and public health planners.
- E is for Envision. Envision a solution and then design it.
- D is for Deploy. Deploy and test your solution.
- Then ask: Is your solution 4ALL? Does your effort give same time access to everyone?

The Prepared4ALL strategies have been used to create collaborations between local emergency management, a public health department, businesses and a university to establish accessible COVID-19 vaccine sites for people who are Deaf or Hard of Hearing. A state public health department used Prepared4ALL concepts to assess COVID-10 vaccine and testing sites for accessibility and to create sensory materials for use in emergency shelters by people with disabilities who also have sensory issues. A volunteer first responder organization has incorporated Prepared4All into its training and Prepared4ALL tools have been successfully used to assess inclusion and equity-related gaps in a county emergency plan.

The University of Sydney's Disability-Inclusive Disaster Risk Reduction (DIDRR) Framework and Toolkit represents another example of a framework and toolkit designed to guide disability-inclusive

FIG. 1

Prepared4ALL strategies [29].

Credit: Association of University Centers on Disabilities (AUCD). Used with permission.

planning while engaging people with disabilities in the planning process [30]. Used not only in Australia, but also throughout Asia, this tool gives a "roadmap" of how people with disabilities, local emergency management, and community based organizations, including those that serve people with disabilities, can come together to design a more disability-inclusive and resilient approach to disasters. The tool provides a number of elements including tips for helping groups begin to interact and work together as well as a companion resource guide providing quick access to the tools. The goal of these tools is to foster quality interactions among the groups so that they can innovate and create ways to address the needs of all populations during a disaster and reduce community disaster risk. The creation of the toolkit involved a similar process, bringing together more than 250 collaborators from multiple sectors. Methods of gathering information included consultations and interviews with stakeholders and key informants in addition to review of relevant literature and documents. Through an iterative process, the group synthesized the gathered information to design and develop the framework and toolkit. The important underlying principles about DIDRR determined from this work appear in Fig. 2.

These three principles guide the work and lead to the eight practice tips included in the toolkit. These practice tips contain valuable information structured using the Australian four-step emergency management cycle: preparedness, response, recovery, prevention [31]. The framework uses the Queensland resilience principles: anticipate, respond, and adapt. It is organized into three sections: impact, outcomes, and processes, with four process approaches to lead to more effective inclusion [30]. The impact of inclusion is to both reduce disaster risk and build disaster resilience within the community, including people with disabilities and the entities that support them. The outcomes of the framework and toolkit seek to clarify and set out the roles and responsibilities of all of the parties including emergency management personnel, people with disabilities, and the community entities that support people with disabilities regarding disaster risk reduction, response and recovery, and improving disaster resilience. Processes incorporate person-centered planning based on the individual's specific functional needs and their formal and informal support network in order to identify barriers and solutions to implement in the event of an emergency. In addition, these planning conversations incorporate the needs of people with disabilities into that of the whole community and include potential partners at the table. One strategy used to facilitate planning conversations is learning and building capacity for both emergency management and people with disabilities which includes understanding functional needs and related barriers in order to collaborate to build better solutions. Using data and evidence about what people with disabilities need is another mechanism that helps support planning conversations to assure that the appropriate resources are available. The third strategy employs strength-based thinking to identify capabilities of

Principle 1: Disability-Inclusive Disaster Risk Reduction (DIDRR) is a human rights issue.

Principle 2: DIDRR actions should be tailored to meet the functional needs of people with

disabilities in a disaster.

Principle 3: Preparedness is not a one-time event, but an on-going process.

FIG. 2

University of Sydney disability-inclusive disaster risk reduction framework and toolkit principles.
Adapted from the toolkit Villeneuve M, Dwine B, Moss M, Abson L, Pertiwi P. Disability inclusive disaster risk reduction (DIDRR) framework and toolkit. A report produced as part of the disability inclusive and disaster resilient Queensland project series. The Centre for Disability Research and Policy. Sydney, NSW: The University of Sydney; 2019 [2006].

1. Person-centered emergency preparedness

2. Invite partners to the planning conversation

3. Collaboration involves learning together

4. Disability data and evidence helps build resilience

5. Inclusive community engagement

6. Inclusion of organizations that support people with disabilities

7. Participation and representation of people with disabilities

8. Asset-based tools for DIDRR

FIG. 3

Practice tips.

Adapted from Villeneuve M, Dwine B, Moss M, Abson L, Pertiwi P. Disability inclusive disaster risk reduction (DIDRR) framework and toolkit. A report produced as part of the disability inclusive and disaster resilient Queensland project series. The Centre for Disability Research and Policy. Sydney, NSW: The University of Sydney; 2019 [2006].

people with disabilities to reduce their own disaster risk and the role of universal design in helping to strengthen accessibility. The final mechanism to strengthen the planning conversations incorporates participation and representation to create shared planning, learning and decision making. Associated practice tips that help operationalize these concepts are included in the framework as are details about next steps for each of these. The practice tips appear in Fig. 3.

Both the University of Sydney Disability-Inclusive Disaster Risk Reduction (DIDRR) Framework and Toolkit and the Prepared4ALL toolkit provide strategies and tools to use for inclusion of people with disabilities in emergency management. These tools help guide the conversations and enhance the capabilities of the emergency and public health preparedness planners, people with disabilities, and the local community which in turn reduces disaster risk and builds resilience.

Building resilience

Resilience comprises an important element in response to and recovery from disasters and in daily life [32]. Both individual resilience and community resilience contribute to more successful and more rapid recovery and minimize the psychological, social and economic impacts of a disaster. Studies of people with physical disabilities identified that resilience was associated with better function in every area except physical functioning [33]. Resilience promoted higher psychological well-being with less anxiety, depression and persistent post-traumatic stress over time. Resilience constructs include multiple factors involving learned skills such as mindfulness, social networks and capital, optimism and some cultural specific elements. For people with disabilities, it has been shown to be protective both in acute traumas such as those around a disaster and in coping with the daily complications of dealing with a chronic health problem. In addition, resilience has been linked to reducing disaster risk in that building capabilities and empowerment contribute to resilience as well as disaster risk reduction. Consequently, promising practices related to building resilience can contribute to disaster-inclusive emergency planning and response.

People over the age of 60 years make up almost 12% of the world's population [34]. As people age, the prevalence of chronic health conditions increases and older adult have significant levels of functional disability [35]. Improvements in health related primarily to health behaviors such as not smoking and control of high blood pressure have decreased the prevalence of disability in the elderly population compared to previous generations. However, studies in the United States show that older people have two or more times the prevalence of stroke, chronic joint problems, and physical disability than younger adults. Older adults with physical disabilities and difficulties with performing daily life activities require support and assistance to leave their home, including disaster evacuation, making them vulnerable to the impact of disasters. Therefore, the non-governmental organization, Help Age, advocates for building disaster resilience in elders. An initiative in Japan called Ibasho Café represents a promising practice in building elder resilience [36].

As in many countries, Japan has an aging population [36]. Often aging populations are viewed as a liability requiring care and resources without contributing to the community. This assessment leads to marginalization of this population, further leading to social isolation. However, in the recognition that older adults contribute knowledge, wisdom and experiences that are important to a community and that this population must become more resilient, the Ibasho Café project was started. Ibasho means "place" in Japanese and often refers to a place where someone belongs, feels a sense of purpose, and brings one's assets in terms of knowledge and experience. Ibasho Cafés were created within the community where older adults could come and congregate with other older adults. They are run by local community members and the people that participate shape it into what meets their needs. Through interaction, members of these Ibasho Cafés build social bonds and networks, increase their self-efficacy and involvement in community activities, and develop trusting relationships where people help each other. This building of social capital is one of the core principles of resilience and social capital contributes to stronger, more resilient communities that are better able to weather disasters [37]. The important features of the Ibasho Cafés can be found in Fig. 4. These principles reflect many of those that are important to disability-inclusive disaster planning as well as those that lead to individual and community resilience.

1. Older adults are a valuable community asset.

2. Create informal gathering places

3. Community members drive development and implementation

4. Involve all generations in the endeavor

5. Respect local cultures and traditions

6. All may participate in normal community life

7. Communities must be environmentally, economically, and socially sustainable

8. Communities must grow organically and embrace imperfection gracefully

FIG. 4

Principles of Ibasho Cafés.

Adapted Emery M, Flora C. Spiraling-up: mapping community transformation with community capitals framework. Community Dev 2006;37(1):19–35.

Research supports strategies to build resilience for both individuals and communities. Concerns always arise related to the ability of adults to learn and change in order to improve their resilience. This is an important consideration as building individual resilience which represents a part of increasing community resilience so addressing the issue in an effective manner is crucial. Höfler presents evidence to support the ability of adults to enhance and improve their psychological resilience through using the principles of adult education [38]. This evidence shows that building social capital for community resilience does not always translate to enhancing the psychological resilience of an individual, but it is the latter that is protective against mental health consequences of disasters. One example of a program that addresses individual resilience related to disaster risk reduction in order to build community resilience is a program through the Bangladesh Red Crescent Society which focuses on building self-efficacy in individuals as a part of enhancing community capacity [38,39]. This program worked to address the needs of women and children and reduce the impact of cyclones which disproportionately affect these populations. Although an older project that occurred after a cyclone in 1991 killed 140,000 people most of whom were women and children, the principles used to engage and empower the women in these communities resemble those still in use and exemplify the use of education in empowering and building resilience in vulnerable and marginalized populations. The program provided group training sessions for women on leadership and disaster preparedness as well as other topics including practical skills as well as knowledge. In addition, separate sessions targeted awareness of the issues by the men of the community.

Additional evidence that adults can build resilience comes from research by Ong and colleagues that identified some windows of time in middle and later life during which there are opportunities for building individual psychological resilience [40]. Although the mechanisms and etiology of this process are unknown, this provides evidence that participation in psychological resilience building activities can enhance psychological responses to traumas and diminish the persistence of stress reactions. In addition, Höfler notes that psychological dysfunction related to either psychological symptoms or disintegration of social networks after a disaster increases the utilization of health and mental health resources, law enforcement (due to increased criminal behavior), and social services, burdening the community and its resources [38]. She argues that decreasing the likelihood of psychological dysfunction by strengthening psychological resilience protects both the individual and the community. This effect is especially true for high-risk populations and serves to reduce disaster risk. As adult education strives to help adults change beliefs and assumptions that affect how they get their needs met, this strategy might prove effective in changing the elements of beliefs and behavior that impact their resilience. Mishra and colleagues note that an individual's perception of risk influences their experience of stress during disasters [41]. Höfler posits that by using adult education principles and practices to have risk-informed conversations with individuals about disasters can strengthen coping skills [38]. This can be accomplished by providing information about what to do during a disaster as well as increasing coping capacity, leading to decreased mental distress in the event of an emergency. Therefore, the use of adult learning principles to address disaster coping skills offers a mechanism to strengthen the resilience of individuals and also to reduce the potential burden on community resources, thus reducing disaster risk.

Behavioral health intervention

Building psychological resilience represents one aspect of addressing behavioral health complications of disasters. The World Health Organization (WHO) identifies the provision of an array of behavioral health services delivered by trained lay personnel or professionals as crucial for managing psycho-logical responses and distress related to disasters [42]. Disaster mental health treatment differs from standard behavioral health in that it strives to prevent known psychological stress responses to disasters from persisting and becoming more long term issues [43]. In addition to an array of services, disaster mental health treatment needs to be available across the cycle of a disaster from the time the disaster happens to recovery and beyond. For people with disabilities, mental health specialists must also consider the increased prevalence of existing mental health disorders. The range of behavioral health services needed during and in the aftermath of a disaster is broad and includes immediate care and support, professional therapy, and sometimes admission to a hospital [42]. The mental health component of disaster plans often lacks detail and resources, but guidance exists on how to address mental health issues in disasters and how to work with specific populations at higher risk for issues. Two such models illustrate this issue, including the US National Institute of Mental Health (NIMH) publication based on a workshop designed to identify best practices in early intervention in disaster mental health and a model developed by Ballan and Sormanti to address disaster-related trauma in people with intellectual disabilities [44,45].

The workshop resulting in the NIMH publication brought together experts from six countries including the United States, Great Britain and Australia as well as academics, public health experts, medical and military personnel to examine the research and determine best practices in mental health early intervention in disasters [44]. The depth of the research and the caliber of the attendees lend credence to this seminal work that sets the direction for early intervention in disaster mental health. The issues addressed included timing of interventions and follow-up, level of training or licensure for those providing the service, and features of appropriate screening. Although much work in the field has occurred since this workshop, the organization, depth of consideration, acknowledgment of dissent as well as areas that need more research and the collaboration and broad expertise of those involved make it a significant body of work. Conclusions related to interventions include that early, brief, focused psychotherapeutic intervention could reduce distress in survivors and that some cognitive behavioral interventions could reduce the incidence, severity and persistence of acute and post-traumatic stress responses and depression. The publication noted that there are interventions that have no evidence of beneficial effect as well as those that can potentially cause harm. These latter two emphasize the need to plan for evidence-based mental health interventions in advance to assure that effective treatments are used. The workshop also recommended screening for symptoms and follow-up for those with evidence of stress responses, pre-existing mental health disorders, those who have lost friends or family, and those whose exposure to trauma was particularly intense or long. As well, early intervention should be offered to survivors on an as needed basis. The workgroup developed a number of questions for future research in order to provide informed guidance.

The work of Ballan and Somanti related to providing disaster mental health services to people with intellectual disabilities (ID) uses many of the principles of the NIMH workgroup [45]. Their work is also informed by an understanding of mental health disorders, trauma and the cognitive abilities of people with ID. These were used to develop guidance to provide mental health support and services to people with ID during and after a disaster. Working with people with ID after a trauma needs to take into account the burden of trauma and loss experienced by this population. Valenti-Hein and Schwartz noted that almost 50% of people with ID in their sample experienced 10 or more episodes of abuse [46]. An overlap of acute stress reactions and trauma can lead to prolonged symptoms [45]. As well, the manifestation of the initial stress reaction to a disaster may include behaviors that are different from

1. Differentiate pre-crisis characteristics and behaviors from those that have developed post-crisis

2. Balance between helping the person and supporting them to act on their own behalf, recognizing that some people may be more dependent on authority figures because of life experience

3. Use simply worded, open-ended questions to assess cognitive, affective and behavioral states and the severity of the impact of the crisis on the person with ID

4. Be mindful of the potential for suicide and tailor assessments to the unique psychosocial development and functioning needs of adults with ID

5. Collaborate with disability service providers and advocacy groups to identify the availability of resources to support this population within the disaster-affected community

FIG. 5

Steps for crisis intervention for people with intellectual disabilities (ID).

Adapted from Ballan MS, Sormanti M. Trauma, grief and the social model: practice guidelines for working with adults with intellectual disabilities in the wake of disasters. Rev Disabil Stud 2006;2(3). [cited 2022 Jan 16]; Available from: https://www.rdsjournal.org/index.php/journal/article/view/339.

what is seen in the general population. For example, increases in compulsivity and somatic complaints are common. Social withdrawal or difficulties with relationships with others may be seen. In addition, self-injurious behaviors generally not seen in the general population may be prevalent. These behaviors will likely be viewed as problematic but often are not recognized as stress reactions. To support people with ID through such events, Ballan and Sormanti developed guidelines to use to help accurately identify and approach symptoms in order to provide the right level of support and prevent the occurrence or persistence of symptoms post disaster. This work contains suggestions and steps for multiple situations, including early intervention, screening and treatment, and strategies for inclusion and resilience building. Their steps for crisis intervention appear in Fig. 5.

Conclusion

Promising practices in disability-inclusive disaster management offer a window into the possible ways to build a more inclusive process. Structural, regulatory and policy changes, inclusion of people with disabilities and others with access and functional needs (including self-advocacy and empowerment), behavioral health, and resilience building represent four areas of disaster management with promising practices leading toward the goal of full community inclusion in disaster management and planning as well as full inclusion in the community in general. The promising practices in these four areas require further dissemination and testing before they reach the level of best practices, but each of them offers a window into building more inclusive communities, disaster risk reduction for people with disabilities, and enhanced community and individual resilience. Each project and its impact offer both positive and negative lessons that can be applied to future endeavors. Concepts of universal design, participation and engagement, roles and responsibilities, and elements of planning that promote resilience in populations of people with disabilities can guide emergency management and public health in practices that will reduce the disparate impacts of disasters on people with disabilities, improve their recovery from disasters, and strengthen their resilience.

Questions for thought

(1) Create a list common elements or themes among the promising practices that makes each practice promising. Find an After Action Report or After Action Report and Improvement Plan online that assesses a local response to an emergency, disaster or pandemic. Use your list as a "scorecard" to identify which of the elements in your list are addressed in the After Action Report.

(2) Browse the Community Toolbox website https://ctb.ku.edu/en/about. The website is a free online resource with tools and toolkits to effect social change. Choose 1 toolkit or tool to review. Apply the toolkit or tool to inclusion in emergency management and public health preparedness. How could you use the toolkit or tool to increase disability-inclusive practices throughout the emergency management cycle?

(3) This chapter discussed promising practices related to roles and responsibilities in the emergency management and public health preparedness systems, emergency plan content, increasing stakeholder participation, and policy change, all to increase inclusive emergency planning and response. Choose one of these promising practice areas. What do you see as the barriers and facilitating factors to making successful changes in that area?

References

[1] Center for Community Health and Development at the University of Kansas. Community Tool Box, 2018. Available from: https://ctb.ku.edu/en.

[2] UNISDR. Sendai framework for disaster risk reduction, 2015. Available from: https://www.unisdr.org/files/43291_sendaiframeworkfordrren.pdf.

[3] Un.org., 2017. Available from: https://www.un.org/disabilities/documents/maps/enablemap.jpg.

[4] GFDRR. Disability inclusion in disaster risk management. Available from: https://www.gfdrr.org/en/publication/disability-inclusion-disaster-risk-management-0.

[5] Ledgerwood D. THE TATTs PROGRAM Institutionalizing disaster preparedness and management capacity of BPBDs in Indonesia through Semi-annual report, 2017. [cited 2022 Jan 17]. Available from: https://pdf.usaid.gov/pdf_docs/PA00SZX1.pdf.

[6] Arbeiter-Samariter-Bund. We help here and now. [cited 2022 Jan 16]. Available from: https://www.asb.de/en.

[7] Ada.gov. 2010 ADA regulations, 2010. Available from: https://www.ada.gov/2010_regs.htm.

[8] LA County Department of Public Health. Strategies for inclusive planning in emergency response. [cited 2022 Jan 16]. Available from: http://publichealth.lacounty.gov/eprp/documents/Strategies.

[9] Kailes JI. Checklist for integrating people with disabilities and others with access and functional needs into emergency planning, response & recovery. 2nd ed. Harris Family Center for Disability and Health Policy; 2014. www.hfcdhp.org.

[10] iaem.org. Access & inclusion. [cited 2022 Jan 16]. Available from: https://www.iaem.org/groups/us-caucuses/access-and-inclusion.

[11] Fox MH, White GW, Rooney C, Rowland JL. Disaster preparedness and response for persons with mobility impairments. J Disabil Policy Stud 2007;17(4):196–205.

[12] Wolf-Fordham S. More seats at the table: Civic engagement in local emergency and public health preparedness planning. Suffolk University; 2019.

[13] Musa GJ, Chiang P-H, Sylk T, Bavley R, Keating W, Lakew B, et al. Use of GIS mapping as a public health tool—from cholera to cancer. Health Serv Insights 2013;6, HSI.S10471. Available from: https://www.ncbi.nlm.nih.gov/pmc/articles/PMC4089751/.

[14] DHHS. Medicare electricity-dependent populations by geography, HHS emPOWER Map 3.0. Available from: https://empowermap.hhs.gov/.

[15] United Nations Enable. Disability and the Japan and Ecuador earthquakes. Available from: https://www.un.org/development/desa/disabilities/news/dspd/disability-the-japan-and-ecuador-earthquakes.html.

[16] UNDRR. Sendai framework for disaster risk reduction 2015-2030, 2015. Available from: https://www.undrr.org/publication/sendai-framework-disaster-risk-reduction-2015-2030.

[17] Uzair Y, Balog-Way S, Koistinen M. Integrating disability inclusion in disaster risk management: The whys and hows. [cited 2022 Jan 16]. Available from: https://blogs.worldbank.org/sustainablecities/integrating-disability-inclusion-disaster-risk-management-whys-and-hows.

[18] Ward M. Included by design: a case for regulation for accessible housing in Australia. In: Cowled CJL, editor. Proceedings of the first international conference on engineering, designing and developing the built environment for sustainable wellbeing. Australia: Queensland University of Technology; 2011. p. 31–5.

[19] NCD.gov. Effective emergency management: Making improvements for communities and people with disabilities, 2009. [cited 2022 Jan 16]. Available from: https://ncd.gov/publications/2009/Aug122009.

[20] Charlton JI. Nothing about us without us: Disability oppression and empowerment. Brantford, ON: W. Ross Macdonald School Resource Services Library; 2011.

[21] WASILC Resource Library. General protocols for CIEP stand-up. Olympia, WA: Washington State Independent Living Council; 2021. [cited 2022 Jan 16]. Available from: https://www.wasilc.org/resource-library.

[22] washingtoncommunitylivingconnections.org. Washington state community living connections. [cited 2022 Jan 16]. Available from: https://washingtoncommunitylivingconnections.org/consite/connect/quick_links/centers_for_independent_living.php.

[23] ACL Administration for Community Living. Centers for independent living, 2014. Available from: https://acl.gov/programs/aging-and-disability-networks/centers-independent-living.

[24] WASILC Resource Library. White paper: Building a case for an D/AFN coordinator at emergency management division. Olympia, WA: Washington State Independent Living Council; 2021. [cited 2022 Jan 16]. Available from: https://www.wasilc.org/resource-library.

[25] Association of University Centers on Disabilities (AUCD) National Center on Disability in Public Health. Prepard4ALL, National technical assistance and training center on disability inclusion in emergency preparedness, https://nationalcenterdph.org/our-focus-areas/emergency-preparedness/prepared4all/.

[26] AUCD's National Center on Disability in Public Health. Prepared4ALL Surveys local emergency managers & public health preparedness staff, 2021. [cited 2022 Jan 16]. Available from: https://nationalcenterdph.org/our-focus-areas/emergency-preparedness/prepared4all/prepared4all-surveys-local-emergency-managers-public-health-preparedness-staff/.

[27] Advocacy Monitor. National council on independent living. Best practices in building relationships in emergency management, 2016. [cited 2022 Jan 16]. Available from: https://advocacymonitor.com/best-practices-in-building-relationships-in-emergency-management/.

[28] FEMA. A whole community approach to emergency management: Principles, themes, and pathways for action, 2011. Available from: https://www.fema.gov/sites/default/files/2020-07/whole_community_dec2011__2.pdf.

[29] Wolf-Fordham S, Griffen A, Owen L, Augustin D, Singleton P. Prepared4ALL supporting disability organizations to engage with local emergency and public health preparedness planners. [Video and PowerPoint Presentation]. In: Presented to the American Public Health association annual meeting; 2021.

[30] Villeneuve M, Dwine B, Moss M, Abson L, Pertiwi P. Disability inclusive disaster risk reduction (DIDRR) framework and toolkit. A report produced as part of the disability inclusive and disaster resilient Queensland project series. Sydney, NSW: The Centre for Disability Research and Policy. The University of Sydney; 2019. 2006.

[31] Australian Disaster Resilience Knowledge Hub. Handbook Australian Emergency Management Arrangements, 2019. Available from: https://knowledge.aidr.org.au/resources/handbook-australian-emergency-management-arrangements/.

[32] Terrill AL, Molton IR, Ehde DM, Amtmann D, Bombardier CH, Smith AE, et al. Resilience, age, and perceived symptoms in persons with long-term physical disabilities. J Health Psychol 2014;21(5):640–9.

[33] Battalio SL, Tang CL, Jensen MP. Resilience and function in adults with chronic physical disabilities: a cross-lagged panel design. Ann Behav Med 2019;54(5):297–307.

[34] Global AgeWatch Index. Global AgeWatch brief 6: Building disaster resilience of older people | Reports, 2015. [cited 2022 Jan 16]. Available from: https://www.helpage.org/global-agewatch/reports/global-agewatch-brief-6-building-disaster-resilience-of-older-people/.

[35] Institute of Medicine. Retooling for an aging America: Building the health care workforce. Washington, DC: National Academies Press; 2008.

[36] Aldrich DP, Kiyota E, Arnold M, Tanaka Y. Elders leading the Way to resilience. SSRN Electron J 2015.

[37] Emery M, Flora C. Spiraling-up: mapping community transformation with community capitals framework. Community Dev 2006;37(1):19–35.

[38] Höfler M. Psychological resilience building in disaster risk reduction: contributions from adult education. Int J Disaster Risk Sci 2014;5(1):33–40.

[39] Schmuck H. Empowering women in Bangladesh—Bangladesh. [cited 2022 Jan 16]. Available from: http://reliefweb.int/report/bangladesh/empowering-women-bangladesh.

[40] Ong A, Bergeman C, Chow SM. Positive emotions as a basic building block in adult resilience. In: Reich J, Zautra A, Hall J, editors. Handbook of adult resilience. New York: Guilford Press; 2010. p. 81–93.

[41] Mishra SK, Suar D, Paton D. Self-esteem and sense of mastery influencing disaster preparedness behaviour. Australas J Disaster Trauma Stud 2011;2011(1).

[42] World Health Organization: WHO. Mental health in emergencies. World Health Organization: WHO; 2019. Available from: https://www.who.int/news-room/fact-sheets/detail/mental-health-in-emergencies.

[43] McIntyre J, Nelson Goff BS. Federal disaster mental health response and compliance with best practices. Community Ment Health J 2011;48(6):723–8.

[44] National Institute of Mental Health. Mental health and mass violence: evidence-based early psychological intervention for victims/survivors of mass violence. In: A Workshop to Reach Consensus on Best Practices. NIH Publication No. 02-5138. Washington, DC: U.S. Government Printing Office; 2002.

[45] Ballan MS, Sormanti M. Trauma, grief and the social model: practice guidelines for working with adults with intellectual disabilities in the wake of disasters. Rev Disabil Stud 2006;2(3). [cited 2022 Jan 16]; Available from: https://www.rdsjournal.org/index.php/journal/article/view/339.

[46] Valenti-Hein D, Schwartz LD. The sexual abuse interview for those with developmental disabilities. Santa Barbara, CA: James Stanfield Co., Inc; 1995.

Index

Note: Page numbers followed by *f* indicate figures, *t* indicate tables, and *b* indicate boxes.

Printed in the United States
by Baker & Taylor Publisher Services